INTERNATIONAL SERIES OF MONOGRAPHS IN
NATURAL PHILOSOPHY
GENERAL EDITOR: D. TER HAAR

VOLUME 65

INTRODUCTION TO FEYNMAN DIAGRAMS

INTRODUCTION TO FEYNMAN DIAGRAMS

by

S. M. BILENKY

Joint Institute for Nuclear Research,
Academy of Sciences of the Ukrainian SSR

Translated and edited by

FRANCIS PARDEE

Institute of Theoretical Science,
University of Oregon, U.S.A.

PERGAMON PRESS

OXFORD · NEW YORK
TORONTO · SYDNEY · BRAUNSCHWEIG

Pergamon Press Ltd., Headington Hill Hall, Oxford
Pergamon Press Inc., Maxwell House, Fairview Park, Elmsford, New York 10523
Pergamon of Canada Ltd., 207 Queen's Quay West, Toronto 1
Pergamon Press (Aust.) Pty. Ltd., 19a Boundary Street, Rushcutters Bay, N.S.W. 2011, Australia
Vieweg & Sohn GmbH, Burgplatz 1, Braunschweig

First edition 1974

Library of Congress Cataloging in Publication Data

Bilen'kiĭ, Samoil Mikhelevich.
Introduction to Feynman diagrams.

(International series of monographs in natural philosophy, v. 65)
Translation of Vvedenie v diagrammnuiû tekhniku Feĭnmana.
Includes bibliographical references.
1. Feynman diagrams. I. Title.
QC794.6.F4B5413 539.7'21 73-21657
ISBN 0-08-017799-9

Printed in Hungary

CONTENTS

FOREWORD

THIS book is based on lectures given to experimental physicists at the Joint Institute for Nuclear Research and students at the Dubna branch of Moscow State University. The lecture style in so far as possible has been preserved and it is for this reason that there are repetitions in the text.

The book presents Feynman diagram techniques and methods for calculating quantities measured experimentally (cross sections, polarizations). It is primarily intended for experimental physicists. The author hopes that the book will also be useful to advanced students specializing in elementary particle physics.

The graphical method of representing a perturbation theory series (the Feynman diagram) which appeared at the beginning of the 1950s has turned out to be extremely fruitful and is firmly established in all areas of physics. The significance of this method goes far beyond the boundaries of perturbation theory.

At the present time Feynman diagrams have become a widespread language which not only theoretical but also experimental physicists should know. Feynman diagrams are very intuitive; the rules for constructing them are simple. However, for a conscientious use of the Feynman diagram method we must proceed by introducing quantized fields, the S-matrix, and chronological and normal products.

The author's goal was to present all these topics as economically as possible. Proofs therefore have been simplified and only the minimum amount of rigor necessary for comprehension has been maintained.

There are examples of the most varied processes in which particles as well as antiparticles, identical particles, and so forth take part. The author tried to present methods for calculating the matrix elements of these processes (by perturbation theory) and the basic rules for constructing Feynman diagrams.

Finally, the author tried to present in detail the methods for calculating quantities measured experimentally. The techniques for calculating cross sections and polarizations is illustrated for a whole series of processes, each of which is examined thoroughly. The details of the calculations can easily be retraced by the reader.

The examples discussed here relate to elementary particle physics and include a wide range of processes with weak, electromagnetic, and strong interactions.

The processes in which both leptons and hadrons take part ($\nu_\mu + n \to \mu^- + p, e + p \to e + p$, and others) are examined in most detail. The electromagnetic and weak form factors of nucleons are discussed in great detail. The author, however, does not pretend to have provided any kind of orderly presentation of the present theory of weak and electromagnetic interactions.

A whole series of important problems in elementary particle physics are not examined in this book. We do not discuss here the analytic characteristics of the matrix elements (dispersion relations), higher orders of perturbation theory and renormalization theory, and much else. The presentation of these problems can be found in books on quantum field theory and elementary particle physics. These books contain further references to the original literature.

In conclusion I consider it my pleasant duty to thank Ya. A. Smorodinsky for useful discussions of the questions examined in the book, as well as D. Fakirov, N. M. Shumeyko, and especially L. L. Nemenov, who read the manuscript and made a number of significant comments.

INTRODUCTION

BEFORE proceeding to the presentation of field theory we shall introduce the system of units generally accepted in relativistic quantum theory, in which Planck's constant $\hbar$ and the speed of light c are equal to one. Let us examine the relationship between this system and the cgs system. In the cgs system $\hbar$ and c are equal respectively to

$$\hbar = (\hbar)\,\frac{\text{g}\times\text{cm}^2}{\text{sec}}, \qquad c = (c)\,\frac{\text{cm}}{\text{sec}},$$

where $(\hbar) \approx 1.05\times 10^{-27}$ and $(c) \approx 3\times 10^{10}$—the values for Planck's constant and the speed of light in the cgs system. Let us choose units of time and mass such that in the new system of units the numerical values of Planck's constant $\hbar$ and the speed of light c are equal to one. If we choose as units of time and mass

$$t_0 = \frac{\text{sec}}{(c)}, \qquad m_0 = \frac{(\hbar)}{(c)}\,\text{g},$$

then obviously in these units

$$\hbar = 1\,\frac{m_0\times\text{cm}^2}{t_0}, \qquad c = 1\,\frac{\text{cm}}{t_0}.$$

It is easy to obtain the relationships between the values of physical quantities in the cgs system and the new system of units. Denote the mass by m, the energy by E, and the momentum by p. Thus

$$m = (m)\,\text{g} = (m)\frac{c}{\hbar}\,m_0 = (m)'\,m_0,$$

$$E = (E)\,\frac{\text{g}\times\text{cm}^2}{\text{sec}^2} = (E)\,\frac{1}{(\hbar)\,(c)}\,\frac{m_0\times\text{cm}^2}{t_0^2} = (E)'\,\frac{m_0\times\text{cm}}{t_0},$$

$$p = (p)\,\frac{\text{g}\times\text{cm}}{\text{sec}} = (p)\,\frac{1}{\hbar}\,\frac{m_0\times\text{cm}}{t_0} = (p)'\,\frac{m_0\times\text{cm}}{t_0}.$$

Here (m), (E), and (p) are the values for m, E, and p in the cgs system and $(m)'$, $(E)'$, and (p) are the values for the corresponding quantities in the system $\hbar = c = 1$. Thus we find

$$(m)' = (m)\,\frac{(c)}{(\hbar)}, \qquad (E)' = (E)\,\frac{1}{(\hbar)\,(c)}, \qquad (p)' = (p)\,\frac{1}{(\hbar)}.$$

In an analogous manner it is not difficult to obtain the relationships between the values of any physical quantities in the cgs system and in the system in which $\hbar = c = 1$.

Not all relationships between physical quantities in the new units contain the dimensional constants $\hbar$ and c (for example, the relativistic relationship between energy and momentum in these units takes the form $E^2 = m^2 + p^2$). This is equivalent to the fact that in the system where $\hbar = c = 1$, $TL^{-1} = 1$ and $ML^2T = 1$ and consequently the units of measurement of all quantities are expressed in terms of L. Obviously the mass has the dimension $M = L^{-2}T = L^{-1}$, the energy has the dimension L^{-1}, the angular momentum is a dimensionless quantity, the momentum has dimension $MLT^{-1} = L^{-1}$, and so forth.

Note that four-vectors will always be written in the form $A = (\mathbf{A}, iA_0)$. The square of this four-vector (a scalar) is $A^2 = \mathbf{A}^2 + A_4^2 = \mathbf{A}^2 - A_0^2$.

CHAPTER 1

THE S-MATRIX

1. The Interaction Representation. The S-matrix

Quantum field theory describes those particle interactions in which the number of particles may not be conserved. We begin by introducing the S-matrix, the matrix whose elements give the probability amplitudes for the corresponding transitions.

The basic postulate of quantum field theory is that the equation of motion is the Schrödinger equation:

$$i\frac{\partial \Psi(t)}{\partial t} = H\Psi(t). \tag{1.1}$$

Here $\Psi(t)$ is the wave function describing the system at time t and H is the full Hamiltonian of the system. If the system is closed, the Hamiltonian does not depend on time. It is further postulated that operators correspond to physical quantities in quantum field theory as in ordinary quantum mechanics and the mean values of the operators

$$\langle O \rangle = (\Psi^+ O \Psi) \tag{1.2}$$

are the observed quantities. In (1.2) O is the operator corresponding to some physical quantity and Ψ is the wave function describing the state.

Let V be an arbitrary unitary operator:

$$V^+ V = 1. \tag{1.3}$$

Then it is obvious that

$$\langle O \rangle = (\Psi'^+ O' \Psi'), \tag{1.4}$$

where†

$$\Psi' = V\Psi, \quad \Psi'^+ = \Psi^+ V^+, \quad O' = VOV^+. \tag{1.5}$$

† We use matrix notation for the operators and functions. In components, equation (1.1) is written in the following manner:

$$i\frac{\partial \Psi_\alpha(t)}{\partial t} = \sum_{\alpha'} H_{\alpha\alpha'}\Psi_{\alpha'}(t).$$

The mean value of the operator O by definition is equal to $\langle O \rangle = \sum_{\alpha', \alpha} \Psi^*_{\alpha'} O_{\alpha'\alpha} \Psi_\alpha$. As a matrix, this expression can be written in the form (1.2). Obviously

$$\Psi'^*_\alpha = \sum_{\alpha'} V^*_{\alpha\alpha'}\Psi^*_{\alpha'} = \sum_{\alpha'} \Psi^*_{\alpha'}(V^+)_{\alpha'\alpha} = (\Psi^+ V^+)_\alpha.$$

Relationships (1.4) and (1.5) imply that the mean value of the operator does not change if we use the wave function $\Psi' = V\Psi$ for describing the system instead of the wave function Ψ, while instead of the operator O, the operator $O' = VOV^+$ corresponds to the physical quantity. In other words, the mean values of the operators (quantities measured experimentally) are invariant under the unitary transformation (1.5).

The unitary operator V can be chosen such that the new wave functions change with time only when there is an interaction. Actually, V is an arbitrary unitary operator. It can also depend on time. Let us denote such an operator by $V(t)$ and the new functions and operators by

$$\Phi(t) = V(t)\,\Psi(t), \quad O(t) = V(t)OV^+(t). \tag{1.6}$$

We write the full Hamiltonian H in the form

$$H = H_0 + H_I, \tag{1.7}$$

where H_0 is the free Hamiltonian and H_I is the interaction Hamiltonian. Substituting $\Psi(t) = V^+(t)\Phi(t)$ in the equation of motion (1.1) we obtain

$$i\frac{\partial V^+(t)}{\partial t}\Phi(t) + iV^+(t)\frac{\partial \Phi(t)}{\partial t} = (H_0 + H_I)\,V^+(t)\,\Phi(t). \tag{1.8}$$

Now let us choose the unitary operator $V(t)$ such that

$$i\frac{\partial V^+(t)}{\partial t} = H_0 V^+(t). \tag{1.9}$$

Multiplying (1.8) by $V(t)$ and using (1.9) we obtain the equation

$$i\frac{\partial \Phi(t)}{\partial t} = H_I(t)\,\Phi(t), \tag{1.10}$$

where

$$H_I(t) = V(t)\,H_I V^+(t). \tag{1.11}$$

The transition from the description of the system with wave functions $\Psi(t)$ which satisfy Schrödinger's equation (1.1) to the description with functions $\Phi(t) = V(t)\,\Psi(t)$, where $V(t)$ is a unitary operator satisfying equation (1.9), is called the transition from the Schrödinger representation to the Dirac representation (interaction representation). The wave function of the system in the interaction representation, as can be seen from equation (1.10), depends on time only when there is an interaction. The interaction representation is widely used in quantum field theory and henceforth we, as a rule, shall work in this representation. Note that the general solution of equation (1.9) is

$$V^+(t) = e^{-iH_0(t-t_1)}V^+(t_1), \tag{1.12}$$

where $V^+(t_1)\,V(t_1) = 1$. If we assume that the representations coincide at time t_1, then $V(t_1) = 1$ and

$$V^+(t) = e^{-iH_0(t-t_1)}. \tag{1.13}$$

As can be seen from (1.6), the operators in the interaction representation depend on time in general. Differentiating $O(t)$ with respect to t and using the equation

$$-i\frac{\partial V(t)}{\partial t} = V(t)\,H_0, \tag{1.14}$$

obtained from (1.9) by Hermitian conjugation, we find

$$i\frac{\partial O(t)}{\partial t} = -V(t)H_0OV^+(t)+V(t)\,OH_0V^+(t) = [O(t), H_0(t)]. \tag{1.15}$$

We assumed for this that the operator O in the original Schrödinger representation does not depend on time. Relationship (1.15) defines the time-dependence of the operators in the interaction representation. From (1.15) it is obvious that the free Hamiltonian in this representation does not depend on time, i.e. $H_0(t) = H_0$.

Now let us determine the general solution of the equation of motion (1.10). For this we examine the equivalent integral equation

$$\Phi(t) = \Phi(t_0)+(-i)\int_{t_0}^{t} dt_1 H_I(t_1)\,\Phi(t_1), \tag{1.16}$$

where $\Phi(t_0)$ is the wave function of the system at initial time t_0. The initial wave function must be given. Our task consists in finding the wave function of the system at any subsequent time t. Substituting for $\Phi(t_1)$ under the integral sign on the right-hand side of (1.16) the sum

$$\Phi(t_0)+(-i)\int_{t_0}^{t_1} dt_2 H_I(t_2)\,\Phi(t_2),$$

we obtain

$$\Phi(t) = \Phi(t_0)+(-i)\int_{t_0}^{t} dt_1 H_I(t_1)\,\Phi(t_0)+(-i)^2\int_{t_0}^{t} dt_1 H_I(t_1)\int_{t_0}^{t_1} dt_2 H_I(t_2)\,\Phi(t_2). \tag{1.17}$$

Continuing this procedure we find

$$\Phi(t) = \Bigg[1+(-i)\int_{t_0}^{t} dt_1 H_I(t_1)+(-i)^2\int_{t_0}^{t} dt_1 H_I(t_1)\int_{t_0}^{t_1} dt_2 H_I(t_2)+\dots$$
$$+(-i)^n\int_{t_0}^{t} dt_1 H_I(t_1)\int_{t_0}^{t_1} dt_2 H_I(t_2)\dots\int_{t_0}^{t_{n-1}} dt_n H_I(t_n)+\dots\Bigg]\Phi(t_0). \tag{1.18}$$

Thus, the general solution of the equation of motion (1.10) is obtained in the form of a series in powers of the interaction Hamiltonian. We are not going to discuss the question of the convergence of this series or the possibility of summing it. We shall only note that in the case when the interaction is characterized by a small constant, the first few members of the series (1.18) already can give the solution with sufficient accuracy. This occurs in the case of the electromagnetic interaction as a result of the smallness of the fine structure constant $\alpha = e^2/4\pi \approx 1/137$.

Let us write the solution (1.18) in the form

$$\Phi(t) = U(t,\ t_0)\,\Phi(t_0), \tag{1.19}$$

where the operator $U(t, t_0)$ is equal to

$$U(t, t_0) = \sum_{n=0}^{\infty} (-i)^n \int_{t_0}^{t} dt_1 H_I(t_1) \int_{t_0}^{t_1} dt_2 H_I(t_2) \dots \int_{t_0}^{t_{n-1}} dt_n H_I(t_n). \tag{1.20}$$

Thus, the solution of the equation of motion at time t can be obtained by operating with $U(t, t_0)$, which is defined by the interaction Hamiltonian, on the wave function given at initial time t_0.

It is not difficult to see that $U(t, t_0)$ is a unitary operator. Actually, it follows from (1.10) and (1.19) that the operator $U(t, t_0)$ satisfies the equation

$$i \frac{\partial U(t, t_0)}{\partial t} = H_I(t)\, U(t, t_0). \tag{1.21}$$

Hence, with the help of the Hermitian conjugate we have (H_I is a Hermitian operator)

$$-i \frac{\partial U^+(t, t_0)}{\partial t} = U^+(t, t_0) H_I(t). \tag{1.22}$$

Let us multiply the left-hand side of (1.21) by $U^+(t, t_0)$ and the right-hand side of (1.22) by $U(t, t_0)$ and subtract the second relationship from the first. We obtain

$$\frac{\partial}{\partial t} (U^+(t, t_0)\, U(t, t_0)) = 0. \tag{1.23}$$

Thus, the operator $U^+(t, t_0)\, U(t, t_0)$ does not depend on t. Obviously

$$U(t_0, t_0) = 1. \tag{1.24}$$

From (1.23) and (1.24) we conclude that

$$U^+(t, t_0)\, U(t, t_0) = 1. \tag{1.25}$$

Let us write the members of the series (1.20) in a more convenient form. We shall first examine the third member of the series. The expression $H_I(t_1)\, H_I(t_2)$ is integrated over the shaded area in Fig. 1. The integration is first carried out over t_2 from t_0 to t_1, and then over t_1 from t_0 to t. Let us integrate $H_I(t_1)\, H_I(t_2)$ over the same area, but first over t_1 (from t_2 to t) and then over t_2 (from t_0 to t). Assuming that the value of the integral does not depend on the order of integration, we find

$$\int_{t_0}^{t} dt_1 \int_{t_0}^{t_1} dt_2\, H_I(t_1)\, H_I(t_2) = \int_{t_0}^{t} dt_2 \int_{t_2}^{t} dt_1\, H_I(t_1)\, H_I(t_2) = \int_{t_0}^{t} dt_1 \int_{t_1}^{t} dt_2\, H_I(t_2)\, H_I(t_1). \tag{1.26}$$

The last equation is obtained by a change of the integration variables ($t_1 \rightleftarrows t_2$). Thus, we have

$$\int_{t_0}^{t} dt_1 \int_{t_0}^{t_1} dt_2\, H_I(t_1)\, H_I(t_2) = \tfrac{1}{2} \int_{t_0}^{t} dt_1 \left[\int_{t_0}^{t_1} dt_2\, H_I(t_1)\, H_I(t_2) + \int_{t_1}^{t} dt_2\, H_I(t_2)\, H_I(t_1) \right]. \tag{1.27}$$

This expression can be written more compactly. For this we introduce the *Dyson chronolog-*

ical operator P:

$$P(H_I(t_1)\,H_I(t_2)) = \begin{cases} H_I(t_1)\,H_I(t_2), & t_1 > t_2, \\ H_I(t_2)\,H_I(t_1), & t_2 > t_1. \end{cases} \tag{1.28}$$

The operator P acting on a product of time-dependent operators arranges them so that the time argument decreases from left to right. From (1.27) and (1.28) we obtain

$$\int_{t_0}^{t} dt_1 \int_{t_0}^{t_1} dt_2\, H_I(t_1)\, H_I(t_2) = \tfrac{1}{2} \int_{t_1}^{t} dt_1 \int_{t_0}^{t} dt_2\, P(H_I(t_1)\, H_I(t_2)). \tag{1.29}$$

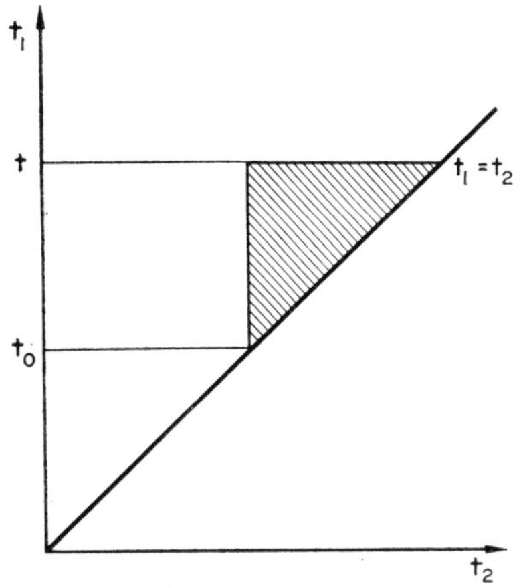

FIG. 1

We shall prove that in the general case

$$\int_{t_0}^{t} dt_1 \int_{t_0}^{t_1} dt_2 \ldots \int_{t_0}^{t_{n-1}} dt_n\, H_I(t_1)\, H_I(t_2) \ldots H_I(t_n)$$
$$= \frac{1}{n!} \int_{t}^{t} dt_1 \int_{t_0}^{t} dt_2 \int_{t_0}^{t} dt_n\, P(H_I(t_1)\, H_I(t_2) \ldots H_I(t_n)), \tag{1.30}$$

where the Dyson chronological operator P in the case of n factors is defined such that

$P(H_I(t_1)\, H_I(t_2) \ldots H_I(t_n)) = H_I(t_1)\, H_I(t_2) \ldots H_I(t_n)$ where $t_1 > t_2 > \ldots > t_n$. (1.31

The proof is by induction. Let relationship (1.30) be true for n factors. Let us examine the expression

$$I(t) = \int_{0}^{t} dt' \int_{t_0}^{t'} dt_1 \int_{t_0}^{t_1} dt_2 \ldots \int_{t_0}^{t_{n-1}} dt_n\, H_I(t')\, H_I(t_1)\, H_I(t_2) \ldots H_I(t_n)$$
$$- \frac{1}{(n+1)!} \int_{t_0}^{t} dt' \int_{t_0}^{t} dt_1 \int_{t_0}^{t} dt_2 \ldots \int_{t_0}^{t} dt_n\, P(H_I(t')\, H_I(t_1)\, H_I(t_2) \ldots H_I(t_n)). \tag{1.32}$$

Differentiating $I(t)$ with respect to t we find

$$\frac{dI(t)}{dt} = H_I(t)\left[\int_{t_0}^{t} dt_1 \int_{t_0}^{t_1} dt_2 \ldots \int_{t_0}^{t_{n-1}} dt_n\, H_I(t_1)\, H_I(t_2) \ldots H_I(t_n) - \frac{1}{n!}\int_{t_0}^{t} dt_1 \int_{t_0}^{t} dt_2 \ldots \int_{t_0}^{t} dt_n\, P(H_I(t_1)\, H_I(t_2) \ldots H_I(t_n))\right]. \tag{1.33}$$

To obtain (1.33) the relationship

$$P(H_I(t_1)\, H_I(t_2) \ldots H_I(t_i)\, H_I(t)\, H_I(t_{i+1}) \ldots H_I(t_n)) = H_I(t)\, P(H_I(t_1)\, H_I(t_2) \ldots H_I(t_n)) \tag{1.34}$$

was used, which is obviously correct for arbitrary n ($t_1 \leqslant t$, $t_2 \leqslant t$, ...). If relationship (1.30) is true, then the right-hand side of (1.33) is equal to zero, i.e. I does not depend on time. Since $I(t_0) = 0$, this implies that $I(t) = 0$ for any t. Thus, from the validity of relationship (1.30) for n factors follows the validity for $n+1$ factors. Since we showed that relationship (1.30) is true for $n = 2$, it is thereby proved in general. Consequently, the operator $U(t, t_0)$ can be written in the form

$$U(t, t_0) = \sum_{n=0}^{\infty} \frac{(-i)^n}{n!} \int_{t_0}^{t} dt_1 \int_{t_0}^{t} dt_2 \ldots \int_{t_0}^{t} dt_n\, P(H_I(t_1) \ldots H_I(t_n)). \tag{1.35}$$

Now we shall formulate the basic problem—the problem of the *collision of particles*. Let us assume that the initial state is given for $t_0 \to -\infty$; we are interested in the state of the system for $t \to \infty$. From (1.19) we obtain

$$\Phi(\infty) = U(\infty, -\infty)\Phi(-\infty). \tag{1.36}$$

Let us denote

$$U(\infty, -\infty) = S. \tag{1.37}$$

This operator is called the *S-matrix*.

Thus, the operator S acting on the initial wave function of the system given as $t \to -\infty$ gives the function of the system as $t \to \infty$. We have

$$S = \sum_{n=0}^{\infty} \frac{(-i)^n}{n!} \int_{-\infty}^{\infty} dt_1 \int_{-\infty}^{\infty} dt_2 \ldots \int_{-\infty}^{\infty} dt_n\, P(H_I(t_1)\, H_I(t_2) \ldots H_I(t_n)). \tag{1.38}$$

From (1.25) and (1.37) it follows that the S-matrix is unitary, i.e.

$$S^+S = 1. \tag{1.39}$$

Let Φ_m be some complete orthonormalized set of functions. Expanding $\Phi(\infty)$ in functions Φ_m we obtain

$$\Phi(\infty) = \sum_m \Phi_m(\Phi_m^+ S\Phi(-\infty)). \tag{1.40}$$

Let us assume that at initial time the system is in the state Φ_n, i.e. $\Phi(-\infty) = \Phi_n$. Then

$$\Phi_n(\infty) = \sum_m \Phi_m(\Phi_m^+ S \Phi_n). \tag{1.41}$$

The matrix element $(\Phi_m^+ S \Phi_n) = S_{m;\,n}$ is therefore the *probability amplitude of the transition* from state Φ_n into state Φ_m. The complete set of functions Φ_m chosen in problems concerning the collision of particles is a set of functions which describe free particles with definite momenta. At the same time, in the general case particles can be different in the initial and final states. Besides that, in the process of colliding new particles may be produced. This type of process is described by quantum field theory. Below we shall give a brief account of the mathematical apparatus of quantum field theory and: (1) construct functions which describe free particles with definite momenta; (2) learn how to calculate (by perturbation theory) the matrix elements for the transition between such states; (3) examine in detail the techniques for calculating transition probabilities.

CHAPTER 2

CLASSICAL FIELDS

2. Equations of Motion

A classical system of N bodies is described by $3N$ functions of time $\mathbf{q}_i(t)$ $(i = 1, \ldots, N)$, which satisfy Newton's equations:

$$m_i \ddot{\mathbf{q}}_i = -\frac{\partial V}{\partial \mathbf{q}_i}. \tag{2.1}$$

Here m_i is the mass of the ith particle and V is the potential energy.

As is known, the equations of motion (2.1) can be obtained from a variational principle. Let us recall its formulation. For the sake of simplicity we shall limit ourselves to the case of one-dimensional motion. Define the action

$$\mathcal{S} = \int_{t_0}^{t_1} L(q, \dot{q})\, dt, \tag{2.2}$$

where the Lagrange function $L(q, \dot{q})$ is equal to

$$L(q, \dot{q}) = \frac{m\dot{q}^2}{2} - V(q). \tag{2.3}$$

Obviously the value of the action $\mathcal{S}$ is determined by the function $q(t)$ over the interval from t_0 to t ($\mathcal{S}$ is a functional of $q(t)$). Besides $q(t)$ let us examine the function

$$q'(t) = q(t) + \delta q(t), \tag{2.4}$$

where $\delta q(t)$ is infinitesimal. Suppose

$$\delta q(t_0) = \delta q(t_1) = 0. \tag{2.5}$$

The increase of the action going from $q(t)$ to $q'(t)$ (the variation of the action) accurate to first order is equal to

$$\delta \mathcal{S} = \int_{t_0}^{t_1} [L(q', \dot{q}') - L(q, \dot{q})]\, dt = \int_{t_0}^{t_1} \left(\frac{\partial L}{\partial q} - \frac{d}{dt} \frac{\partial L}{\partial \dot{q}} \right) \delta q\, dt. \tag{2.6}$$

In obtaining (2.6) we used the fact that

$$\delta\dot{q} = \dot{q}' - \dot{q} = \frac{d}{dt}\,\delta q,$$

then integrated by parts and took (2.5) into account. We require that the variation of the action (2.6) vanish. Since $\delta q(t)$ is arbitrary, from (2.6) we find

$$\frac{\partial L}{\partial q} - \frac{d}{dt}\frac{\partial L}{\partial \dot{q}} = 0. \tag{2.7}$$

Differentiating the Lagrangian function (2.3) with respect to q and $\dot{q}$, we obtain from (2.7)

$$m\ddot{q} = -\frac{\partial V}{\partial q},$$

i.e. (2.7) is the equation of motion. Thus, the function $q(t)$, which is the solution of the equation of motion, provides an extremum of the action [under the condition (2.5)]. Note that with the help of the Lagrange function the energy and momentum of the particle can also be obtained:

$$H = \frac{m\dot{q}^2}{2} + V = \frac{\partial L}{\partial \dot{q}}\,\dot{q} - L, \qquad p = m\dot{q} = \frac{\partial L}{\partial \dot{q}}. \tag{2.8}$$

We now turn to an examination of classical fields. A well-known example is the electromagnetic field which is described by the potential $A_\mu(\mathbf{x}, t)$ ($\mathbf{A}$ is the vector potential, $A_0 = -iA_4$ is the scalar potential). This means that the electromagnetic field is described by an infinite number of functions of time (the values of the potential at all points of space), i.e. the electromagnetic field is a system with an infinite number of degrees of freedom. The equations of the field (Maxwell's equations) can be obtained from the variational principle which is a general principle applicable to any physical system.

We shall examine fields in addition to the electromagnetic field. Denote the functions describing some field by $\psi_\alpha(x)$, where the index α assumes integer values and $x = (\mathbf{x}, ix_0)$, $x_0 \equiv t$. We write the Lagrangian of the field in the form

$$L = \int \mathcal{L}\left(\psi, \frac{\partial \psi}{\partial x}\right) d\mathbf{x}. \tag{2.9}$$

Here $\mathcal{L}(\psi, \partial\psi/\partial x)$ is the Lagrangian of a unit volume (the Lagrangian density).† We assumed that this quantity depends only on the functions $\psi_\alpha(x)$ and their first derivatives $\partial\psi_\alpha/\partial x_\mu$ [compare (2.3)]. Define the action

$$\mathcal{S} = \int_\Omega \mathcal{L}\left(\psi, \frac{\partial \psi}{\partial x}\right) dx, \tag{2.10}$$

where $dx = d\mathbf{x}\, dx_0$ and Ω is some volume of space-time. The variation of the action is

† Henceforth we shall call this quantity the Lagrangian.

equal to

$$\delta\mathcal{S} = \int\limits_{\Omega} \mathcal{L}\left(\psi', \frac{\partial \psi'}{\partial x}\right) dx - \int\limits_{\Omega} \mathcal{L}\left(\psi, \frac{\partial \psi}{\partial x}\right) dx$$

$$= \int\limits_{\Omega} \left(\frac{\partial \mathcal{L}}{\partial \psi_\alpha} - \frac{\partial}{\partial x_\mu} \frac{\partial \mathcal{L}}{\partial \dfrac{\partial \psi_\alpha}{\partial x_\mu}}\right) \delta\psi_\alpha \, dx + \int\limits_{\Omega} \frac{\partial}{\partial x_\mu} \left(\frac{\partial \mathcal{L}}{\partial \dfrac{\partial \psi_\alpha}{\partial x_\mu}} \delta\psi_\alpha\right) dx. \tag{2.11}$$

Here $\psi'_\alpha(x) = \psi_\alpha(x) + \delta\psi_\alpha(x)$ where $\delta\psi_\alpha(x)$ is infinitesimal.† The equations of the field follow from the variational principle

$$\delta\mathcal{S} = 0 \tag{2.12}$$

with the condition that the functions $\delta\psi_\alpha$ vanish at the boundary of the region of integration. Using the Gauss–Ostrogradsky theorem we find

$$\int\limits_{\Omega} \frac{\partial}{\partial x_\mu} \left(\frac{\partial \mathcal{L}}{\partial \dfrac{\partial \psi_\alpha}{\partial x_\mu}} \delta\psi_\alpha\right) dx = \int\limits_{\Sigma} \frac{\partial \mathcal{L}}{\partial \dfrac{\partial \psi_\alpha}{\partial x_\mu}} \delta\psi_\alpha \, d\sigma_\mu, \tag{2.13}$$

where Σ is the surface of the volume Ω and $d\sigma_\mu$ is a four-vector directed along the external normal of the surface Σ. Equating the variation of the action to zero under the condition that $\delta\psi_\alpha = 0$ on the surface Σ, we obtain the equations of the field

$$\frac{\partial \mathcal{L}}{\partial \psi_a} - \frac{\partial}{\partial x_\mu} \frac{\partial \mathcal{L}}{\partial \dfrac{\partial \psi_a}{\partial x_\mu}} = 0. \tag{2.14}$$

Note that in the case when there are several fields described by the functions $\psi_\alpha(x)$, $\phi_\beta(x)$, ..., the Lagrangian depends on the functions ψ_α, ϕ_β, ..., and their first derivatives. From the variational principle along with (2.14) we obtain

$$\frac{\partial \mathcal{L}}{\partial \phi_\beta} - \frac{\partial}{\partial x_\mu} \left(\frac{\partial \mathcal{L}}{\partial \dfrac{\partial \phi_\beta}{\partial x_\mu}}\right) = 0 \tag{2.15}$$

and so forth.

The equations of the field must be invariant under Lorentz transformations. This requirement imposes a limitation on the possible form of the Lagrangian $\mathcal{L}$. Let us examine the homogeneous Lorentz transformation

$$x'_\mu = a_{\mu\nu} x_\nu. \tag{2.16}$$

Here x_ν and x'_μ are coordinates of the same point in two systems of reference and the

† In (2.11) and all further relationships repeated indices mean summation.

coefficients $a_{\mu\nu}$ satisfy the conditions

$$a_{\mu\nu}a_{\mu\nu'} = \delta_{\nu\nu'}, \tag{2.17}$$

which follows from the requirement that

$$x'_\mu x'_\mu = x_\mu x_\mu. \tag{2.18}$$

Multiplying (2.16) by $a_{\mu\nu}$, summing over μ, and using (2.17), we obtain

$$x_\nu = a_{\mu\nu}x'_\mu. \tag{2.19}$$

From (2.18) and (2.19) it follows that the coefficients $a_{\mu\nu}$ also satisfy the conditions

$$a_{\mu\nu}a_{\mu'\nu} = \delta_{\mu\mu'}. \tag{2.20}$$

In matrix form these relationships can be written in the following manner:

$$x' = ax, \tag{2.16a}$$

$$\tilde{a}a = 1, \tag{2.17a}$$

$$x = \tilde{a}x', \tag{2.19a}$$

$$a\tilde{a} = 1, \tag{2.20a}$$

where the sign $\sim$ denotes transposition. From (2.17) it is not difficult to see that

$$\det a = \pm 1. \tag{2.21}$$

The inversion of space axes

$$\mathbf{x}' = -\mathbf{x}, \quad x'_4 = x_4 \tag{2.22}$$

is an example of a transformation with $\det a = -1$.

We shall now examine functions of fields which transform under the coordinate transformation (2.16) either as tensors or as spinors. In quantum theory the quanta of such fields are particles with a definite spin and intrinsic parity. We shall cite examples of the transformation of the simplest functions. Scalar $\phi(x)$, pseudoscalar $\phi_P(x)$, and vector $A_\mu(x)$ functions are transformed in the following manner:

$$\phi'(x') = \phi(x), \tag{2.23}$$

$$\phi'_P(x') = \det a\phi_P(x), \tag{2.24}$$

$$A'_\mu(x') = a_{\mu\nu}A_\nu(x). \tag{2.25}$$

On the left-hand side of these relationships are functions describing fields in the primed reference system. A spinor function is transformed in the following manner:

$$\psi'_\sigma(x') = L_{\sigma\sigma'}\,\psi_{\sigma'}(x), \quad \bar{\psi}'_\sigma(x') = \bar{\psi}_{\sigma'}(x)\,L^{-1}_{\sigma'\sigma} \tag{2.26}$$

or in matrix form

$$\psi'(x') = L\psi(x), \quad \bar{\psi}'(x') = \bar{\psi}(x)\,L^{-1}. \tag{2.26a}$$

Here the indices σ and σ' assume four values, $\bar{\psi} = \psi^+\gamma_4$, and the matrix L satisfies the relationships

$$L^{-1}\gamma_\mu L = a_{\mu\nu}\gamma_\nu \tag{2.27}$$

(γ_μ is a Hermitian 4×4 matrix obeying the anticommutation relationships $\gamma_\mu\gamma_\nu+\gamma_\nu\gamma_\mu = 2\delta_{\mu\nu}$).

Let us now consider what conditions the Lagrangian of the field must obey so that the equations of the field are invariant under a Lorentz transformation. In the integral (2.10) we change variable x to x' and $\psi_\alpha(x)$ to $\psi'_\alpha(x')$. We obtain

$$\mathcal{S} = \int\limits_{\Omega} \mathcal{L}\left(\psi(x), \frac{\partial\psi(x)}{\partial x}\right) dx = \int\limits_{\Omega'} \mathcal{L}'\left((\psi'(x'), \frac{\partial\psi'(x')}{dx'}\right) dx'. \tag{2.28}$$

We require that

$$\mathcal{L}\left(\psi(x), \frac{\partial\psi(x)}{\partial x}\right) = \mathcal{L}'\left(\psi'(x'), \frac{\partial\psi'(x')}{\partial x'}\right) = \mathcal{L}\left(\psi'(x'), \frac{\partial\psi'(x')}{dx'}\right). \tag{2.29}$$

This condition means that the Lagrangians in the original and the primed systems are the same function of the corresponding fields. If (2.29) is satisfied, we obtain from the variation principle in the original system along with equations (2.14) the following equations of the field in the primed reference system:

$$\frac{\partial\mathcal{L}\left(\psi', \frac{\partial\psi'}{\partial x'}\right)}{\partial\psi'_\alpha} - \frac{\partial}{\partial x'_\mu}\frac{\partial\mathcal{L}\left(\psi', \frac{\partial\psi'}{\partial x'}\right)}{\partial\frac{\partial\psi'_\alpha}{\partial x'_\mu}} = 0. \tag{2.30}$$

Obviously the equations of the field in the primed system are obtained from the equations of the original system by exchanging the unprimed quantities for the primed, i.e. the equations of the field are invariant under Lorentz transformations.

We shall now take as an example the electromagnetic field. The equations of the field are Maxwell's equations

$$\frac{\partial F_{\mu\nu}}{\partial x_\nu} = j_\mu, \tag{2.31}$$

where j_μ is the current (in Heaviside units) and the electromagnetic field strength tensor $F_{\mu\nu}$ is equal to

$$F_{\mu\nu} = \frac{\partial A_\nu}{\partial x_\mu} - \frac{\partial A_\mu}{\partial x_\nu}. \tag{2.32}$$

From (2.31), taking into account the additional Lorentz condition

$$\frac{\partial A_\mu}{\partial x_\mu} = 0, \tag{2.33}$$

we obtain the equation for the potential A_μ

$$\Box A_\mu = -j_\mu, \tag{2.34}$$

where

$$\Box = \frac{\partial^2}{\partial x_\nu^2}.$$

On the other hand, the equations of the field are

$$\frac{\partial \mathcal{L}}{\partial A_\mu} - \frac{\partial}{\partial x_\nu} \frac{\partial \mathcal{L}}{\partial \frac{\partial A_\mu}{\partial x_\nu}} = 0. \tag{2.35}$$

We shall construct the Lagrangian $\mathcal{L}$ so that equations (2.35) coincide with the equations of the field (2.31). Obviously the derivatives of the potential must enter the Lagrangian quadratically and the current j_μ in (2.31) results from differentiating the Lagrangian with respect to A_μ. It is not difficult to verify that the Lagrangian

$$\mathcal{L} = -\tfrac{1}{4} F_{\mu\sigma} F_{\mu\sigma} + j_\mu A_\mu \tag{2.36}$$

satisfies all necessary requirements. Actually,

$$\frac{\partial \mathcal{L}}{\partial A_\mu} = j_\mu, \quad \frac{\partial \mathcal{L}}{\partial \frac{\partial A_\mu}{\partial x_\nu}} = \tfrac{1}{2}(F_{\mu\nu} - F_{\nu\mu}) = F_{\mu\nu}, \tag{2.37}$$

and from (2.35) we obtain Maxwell's equations (2.31). Obviously the Lagrangian (2.36) also satisfies the requirement of invariance under Lorentz transformations.

3. Conserved Quantities

We have assumed that the Lagrangian of the field is a function of $\psi_\alpha(x)$, $\partial\psi_\alpha(x)/\partial x_\mu$, and does not explicitly depend on x. In order to understand what this assumption means, let us examine the transformation

$$\left.\begin{aligned} x'_\mu &= x_\mu + a_\mu, \\ \psi'_\alpha(x') &= \psi_\alpha(x), \end{aligned}\right\} \tag{3.1}$$

where a_μ is an arbitrary four-vector. If the Lagrangian does not explicitly depend on x, then obviously it is invariant under the transformations (3.1), i.e.

$$\mathcal{L}\left(\psi(x), \frac{\partial\psi(x)}{\partial x}\right) = \mathcal{L}\left(\psi'(x'), \frac{\partial\psi'(x')}{\partial x'}\right). \tag{3.2}$$

The invariance of the Lagrangian under the transformations (3.1) (translations) is called *translational invariance*. From the translational invariance it follows that the four-vector

$$P_\mu = i \int T_{\mu 4}\, dx, \tag{3.3}$$

where

$$T_{\mu\nu} = \frac{\partial \mathcal{L}}{\partial \frac{\partial \psi_\alpha}{\partial x_\nu}} \frac{\partial \psi_\alpha}{\partial x_\mu} - \mathcal{L}\delta_{\mu\nu}, \tag{3.4}$$

does not depend on time [the integration in (3.3) is performed over all space]. P_μ is the energy-momentum vector.

In order to make sure that P_μ is conserved, we examine the transformation (3.1) with infinitesimal a. From (3.1) we obtain, accurate to terms of first order in a,

$$\psi'_\alpha(x) = \psi_\alpha(x-a) = \psi_\alpha(x) - a_\mu \frac{\partial \psi_\alpha}{\partial x_\mu}. \tag{3.5}$$

Let us write (3.1) and (3.5) in the form

$$\left.\begin{aligned} \psi'_\alpha &= \psi_\alpha(x) + \delta\psi_\alpha(x), \\ x'_\mu &= x_\mu + \delta x_\mu, \end{aligned}\right\} \tag{3.6}$$

where

$$\left.\begin{aligned} \delta\psi_\alpha(x) &= -a_\mu \frac{\partial \psi_\alpha}{\partial x_\mu}, \\ \delta x_\mu &= a_\mu \end{aligned}\right\} \tag{3.7}$$

are infinitesimal quantities. From the condition of translational invariance (3.2), accurate to terms of first order in a, we find

$$\delta\mathcal{L} + \frac{\partial \mathcal{L}}{\partial x_\nu}\,\delta x_\nu = 0. \tag{3.8}$$

Here

$$\delta\mathcal{L} = \mathcal{L}\left(\psi'(x), \frac{\partial\psi'(x)}{\partial x}\right) - \mathcal{L}\left(\psi(x), \frac{\partial\psi(x)}{\partial x}\right). \tag{3.9}$$

Using equation (2.14) we obtain

$$\delta\mathcal{L} = \frac{\partial}{\partial x_\nu}\left(\frac{\partial\mathcal{L}}{\partial\dfrac{\partial\psi_\alpha}{\partial x_\nu}}\,\delta\psi_\alpha\right). \tag{3.10}$$

Thus, from the invariance of the Lagrangian under the transformations (3.6) we find that

$$\frac{\partial}{\partial x_\nu}\left(\frac{\partial\mathcal{L}}{\partial\dfrac{\partial\psi_\alpha}{\partial x_\nu}}\,\delta\psi_\alpha + \mathcal{L}\,\delta x_\nu\right) = 0. \tag{3.11}$$

With the help of (3.11) and (3.7) we obtain

$$\frac{\partial}{\partial x_\nu}\left(\mathcal{L}\delta_{\mu\nu} - \frac{\partial\mathcal{L}}{\partial\dfrac{\partial\psi_\alpha}{\partial x_\nu}}\,\frac{\partial\psi_\alpha}{\partial x_\mu}\right) a_\mu = 0. \tag{3.12}$$

Hence

$$\frac{\partial T_{\mu\nu}}{\partial x_\nu} = 0, \tag{3.13}$$

where the quantities $T_{\mu\nu}$ are defined by relationship (3.4).

The quantities $T_{\mu\nu}$ transform like a second-rank tensor. This tensor is called the energy-momentum tensor of the field. From (3.13) it follows that the four-vector P_μ, defined by relationship (3.3), is conserved. In order to show this, we shall integrate (3.13) over some volume V. Using the Gauss–Ostrogradsky theorem we obtain

$$\frac{1}{i}\frac{\partial}{\partial x_0}\int_V T_{\mu 4}\,dx + \int_S T_{\mu i} ds_i = 0. \tag{3.14}$$

Here S is the surface enclosing the volume V and the expression is summed over i from 1 to 3.† Taking the limit $V \to \infty$ and imposing natural bounding conditions (the absence of fields at infinity), we find that

$$\frac{\partial}{\partial x_0}\int T_{\mu 4}\,dx = 0. \tag{3.15}$$

Thus, the energy-momentum vector P_μ defined by relationship (3.3) does not depend on time. The fourth component of the energy-momentum vector is equal to

$$P_4 = i\int\left(\frac{\partial \mathscr{L}}{\partial \dfrac{\partial \psi_\alpha}{\partial x_4}}\frac{\partial \psi_\alpha}{\partial x_4} - \mathscr{L}\right) dx = iH, \tag{3.16}$$

where H is the energy of the field [cf. expression (2.8)]. For the energy density $\mathscr{H}$ of the field we find

$$\mathscr{H} = T_{44} = \frac{\partial \mathscr{L}}{\partial \dfrac{\partial \psi_\alpha}{\partial x_4}}\frac{\partial \psi_\alpha}{\partial x_4} - \mathscr{L}. \tag{3.17}$$

Let us show that as a consequence of the invariance of the Lagrangian under Lorentz transformations, the angular momentum tensor of the field is conserved. We shall examine Lorentz transformations which differ from the identity transformations by an infinitesimal quantity. The transformation coefficients $a_{\mu\varrho}$ in (2.16) can be represented in this case in the form

$$a_{\mu\varrho} = \delta_{\mu\varrho} + \varepsilon_{\mu\varrho}, \tag{3.18}$$

where the parameters $\varepsilon_{\mu\varrho}$ are infinitesimal quantities. From conditions (2.20), accurate to first order in ε, we obtain

$$\varepsilon_{\mu\varrho} = -\varepsilon_{\varrho\mu}. \tag{3.19}$$

The Lorentz transformation (2.16) can be written in the form

$$x'_\mu = x_\mu + \delta x_\mu, \tag{3.20}$$

where

$$\delta x_\mu = \varepsilon_{\mu\varrho} x_\varrho. \tag{3.21}$$

† Greek indices $\mu, \nu, \ldots$, as a rule, assume the values 1, 2, 3, 4, and Latin $i, k, \ldots$ assume the values 1, 2, 3.

The functions which describe the field are transformed by the Lorentz transformation (3.20) in the following manner:

$$\psi'_4(x') = (\delta_{\alpha\beta}+\tfrac{1}{2}\Sigma^{\mu\varrho}_{\alpha\beta}\varepsilon_{\mu\varrho})\,\psi_\beta(x), \tag{3.22}$$

where as a result of (3.19)

$$\Sigma^{\mu\varrho}_{\alpha\beta} = -\Sigma^{\varrho\mu}_{\alpha\beta}. \tag{3.23}$$

The matrices $\frac{1}{2}\Sigma^{\mu\varrho}$ are the coefficients of $\varepsilon_{\mu\varrho}$ in the expansion of the matrices a, L, and so forth [see (2.25) and (2.26)]. It is clear that for scalar and pseudoscalar fields the matrix $\Sigma^{\mu\varrho}$ is to be set equal to zero. From (3.22), accurate to first order in ε, we find

$$\psi'_\alpha(x) = \psi_\alpha(x)+\delta\psi_\alpha(x), \tag{3.24}$$

where

$$\delta\psi_\alpha(x) = -\frac{\partial\psi_\alpha(x)}{\partial x_\mu}\,\delta x_\mu+\tfrac{1}{2}\Sigma^{\mu\varrho}_{\alpha\beta}\varepsilon_{\mu\varrho}\psi_\beta(x). \tag{3.25}$$

It is obvious that the relationship (3.8), in which δx and $\delta\psi$ are given by relationships (3.21) and (3.25), follows from (2.29)—the condition of invariance of the Lagrangian under Lorentz transformations.

We have

$$\frac{\partial\mathcal{L}}{\partial\dfrac{\partial\psi_\alpha}{\partial x_\nu}}\,\delta\psi_\alpha+\mathcal{L}\delta x_\nu = \left(-T_{\mu\nu}x_\varrho+T_{\varrho\nu}x_\mu+\frac{\partial\mathcal{L}}{\partial\dfrac{\partial\psi_\alpha}{\partial x_\nu}}\,\Sigma^{\mu\varrho}_{\alpha\beta}\psi_\beta\right)\tfrac{1}{2}\,\varepsilon_{\mu\varrho}. \tag{3.26}$$

Here $T_{\mu\nu}$ is the energy-momentum tensor. The angular momentum density tensor is defined by

$$m_{\mu\varrho;\,\nu} = x_\mu T_{\varrho\nu}-x_\varrho T_{\mu\nu}+\frac{\partial\mathcal{L}}{\partial\dfrac{\partial\psi_\alpha}{\partial x_\nu}}\,\Sigma^{\mu\varrho}_{\alpha\beta}\psi_\beta. \tag{3.27}$$

Note that this tensor is antisymmetric under transposition of the indices μ and ϱ:

$$m_{\mu\varrho;\,\nu} = -m_{\varrho\mu;\,\nu}. \tag{3.28}$$

From (3.8), (3.10), and (3.26), recognizing that $\partial\delta x_\nu/\partial x_\nu = 0$, we obtain

$$\frac{\delta m_{\mu\varrho;\,\nu}}{\partial x_\nu} = 0. \tag{3.29}$$

Hence, analogous to (3.14) and (3.15), we find that the tensor

$$M_{\mu\varrho} = i\int m_{\mu\varrho;\,4}\,dx \tag{3.30}$$

does not depend on time. The space components of this tensor are the angular momentum components of the field (M_{12} is the projection of the angular momentum on axis 3 and so forth). Note that

$$L_{\mu\varrho} = i\int (x_\mu T_{\varrho 4}-x_\varrho T_{\mu 4})\,dx \tag{3.31}$$

is the orbital angular momentum and the term

$$S_{\mu\varrho} = i \int \frac{\partial \mathcal{L}}{\partial \dfrac{\partial \psi_\alpha}{\partial x_4}} \Sigma^{\mu\varrho}_{\alpha\beta} \psi_\beta \, dx \tag{3.32}$$

is related to the presence of discrete indices of the functions describing the field and corresponds in quantum field theory to the contribution of particle spins to the total angular momentum.

In conclusion we examine the gauge transformations and the conserved quantities related to the invariance of the Lagrangian under these transformations. Functions which describe fields can be both real and complex. If the functions $\psi_\alpha(x)$, where $\alpha = 1, \ldots, n$, are complex, the field is described by $2n$ real functions—the real and imaginary parts of the functions $\psi_\alpha(x)$. Obviously the real and imaginary parts of the function $\psi_\alpha(x)$ are expressed in terms of $\psi_\alpha(x)$ and $\psi^*_\alpha(x)$. We shall choose as independent dynamic variables the functions $\psi_\alpha(x)$ and $\psi^*_\alpha(x)$, i.e. we shall assume that the Lagrangian depends on $\psi_\alpha(x)$, $\psi^*_\alpha(x)$, and their first derivatives in x_μ. In this case from the variational principle and the equations of the field (2.14) we obtain the equations

$$\frac{\partial \mathcal{L}}{\partial \psi^*_\alpha} - \frac{\partial}{\partial x_\mu} \frac{\partial \mathcal{L}}{\partial \dfrac{\partial \psi^*_\alpha}{\partial x_\mu}} = 0. \tag{3.33}$$

It is also obvious that for the energy-momentum tensor we obtain the following expression:

$$T_{\mu\nu} = \frac{\partial \mathcal{L}}{\partial \dfrac{\partial \psi_\alpha}{\partial x_\nu}} \frac{\partial \psi_\alpha}{\partial x_\mu} + \frac{\partial \mathcal{L}}{\partial \dfrac{\partial \psi^*_\alpha}{\partial x_\nu}} \frac{\partial \psi^*_\alpha}{\partial x_\mu} - \mathcal{L}\, \delta_{\mu\nu}. \tag{3.34}$$

Note that the Lagrangian of the field must be real. Then the energy, momentum, and other physical quantities characterizing the field are real.

Let us examine the transformations

$$\left.\begin{aligned} \psi'_\alpha(x) &= \psi_\alpha(x)\, e^{ie\chi}, \\ \psi^{*\prime}_\alpha(x) &= \psi^*_\alpha(x)\, e^{-ie\chi}. \end{aligned}\right\} \tag{3.35}$$

Here e is a real number which can be different for different fields and χ is an arbitrary real parameter, the same for all fields in the system under consideration. The transformations (3.35) are called gauge transformations. Suppose that the Lagrangian of the field is invariant under the gauge transformations (3.35), i.e.

$$\mathcal{L}\left(\psi, \psi^*, \frac{\partial \psi}{\partial x}, \frac{\partial \psi^*}{\partial x}\right) = \mathcal{L}\left(\psi', \psi^{*\prime}, \frac{\partial \psi'}{\partial x} \frac{\partial \psi^{*\prime}}{\partial x}\right). \tag{3.36}$$

Let us examine the gauge transformations (3.35) with the parameter infinitesimal. Accurate to first order in χ, we obtain

$$\left.\begin{aligned} \psi'_\alpha &= \psi_\alpha + \delta\psi_\alpha, \\ \psi^{*\prime}_\alpha &= \psi^*_\alpha + \delta\psi^*_\alpha, \end{aligned}\right\} \tag{3.37}$$

where

$$\left.\begin{aligned} \delta\psi_\alpha &= ie\chi\psi_\alpha, \\ \delta\psi_\alpha^* &= -ie\chi\psi_\alpha^*. \end{aligned}\right\} \tag{3.38}$$

From the invariance of the Lagrangian under the transformations (3.37), using the equations of the field (2.14) and (3.33), we find

$$\delta\mathscr{L} = \frac{\partial}{\partial x_\mu}\left(\frac{\partial\mathscr{L}}{\partial\dfrac{\partial\psi_\alpha}{\partial x_\mu}}\,\delta\psi_\alpha + \frac{\partial\mathscr{L}}{\partial\dfrac{\partial\psi_\alpha^*}{\partial x_\mu}}\,\delta\psi_\alpha^*\right) = 0. \tag{3.39}$$

With the help of (3.38) and (3.39) we find (χ does not depend on x) that the four-vector

$$j_\mu = -ie\left(\frac{\partial\mathscr{L}}{\partial\dfrac{\partial\psi_\alpha}{\partial x_\mu}}\,\psi_\alpha - \frac{\partial\mathscr{L}}{\partial\dfrac{\partial\psi_\alpha^*}{\partial x_\mu}}\,\psi_\alpha^*\right) \tag{3.40}$$

satisfies the continuity equation

$$\frac{\partial j_\mu}{\partial x_\mu} = 0. \tag{3.41}$$

The vector j_μ, which satisfies equation (3.41) due to gauge invariance, is the current four-vector. Its fourth component is

$$j_4 = i\varrho, \tag{3.42}$$

where ϱ is the charge density. From (3.41) it follows that the charge of the field

$$Q = -i\int j_4\,dx \tag{3.43}$$

(the integration is performed over all space) is a conserved quantity. Quanta of the complex field, as we shall see below, are particles with charge e or $-e$. Note that the gauge transformations (3.35) are inadmissible for fields described by real functions. This means that for real fields $e = 0$. Quanta of such fields are neutral particles.

In order to show that the four-vector j_μ defined by relationship (3.40) is the electrical current, we examine the system described by the complex functions $\psi_\alpha(x)$ and the electromagnetic potential A_μ. We write the Lagrangian of the system in the form

$$\mathscr{L} = \mathscr{L}_0 + \mathscr{L}', \tag{3.44}$$

where

$$\mathscr{L}_0 = -\tfrac{1}{4}F_{\mu\nu}F_{\mu\nu} \tag{3.45}$$

is the Lagrangian of the free electromagnetic field. Let us assume that $\mathscr{L}'$ depends on the functions ψ_α, ψ_α^*, their first derivatives and A_μ, and does not depend on the derivatives of the electromagnetic potential. In place of the potential A_μ one can use the electromagnetic potential

$$A'_\mu = A_\mu + \frac{\partial\Lambda}{\partial x_\mu}, \tag{3.46}$$

where $\Lambda(x)$ is an arbitrary function satisfying the equation

$$\Box\Lambda = 0. \tag{3.47}$$

Here, as is well known, the field strength tensor $F_{\mu\nu}$ does not change and the new potential A'_μ will satisfy the additional Lorentz condition (2.33). If one replaces A_μ by $A'_\mu - (\partial\Lambda/\partial x_\mu)$ in $\mathcal{L}'$, the quantity $\partial\Lambda/\partial x_\mu$, having no physical meaning, must be compensated for by the corresponding changes of the phase of the functions ψ_α and ψ^*_α. The latter can occur only in the case where the potential A_μ enters $\mathcal{L}'$ in the following combinations:

$$\left(\frac{\partial}{\partial x_\mu} - ieA_\mu\right)\psi_\alpha \quad \text{and} \quad \left(\frac{\partial}{\partial x_\mu} + ieA_\mu\right)\psi^*_\alpha. \tag{3.48}$$

Actually,

$$\left(\frac{\partial}{\partial x_\mu} - ieA_\mu\right)\psi_\alpha = \left(\frac{\partial}{\partial x_\mu} - ieA'_\mu + ie\frac{\partial\Lambda}{\partial x_\mu}\right)\psi_\alpha = e^{-ie\Lambda}\left(\frac{\partial}{\partial x_\mu} - ieA'_\mu\right)\psi'_\alpha. \tag{3.49}$$

Here

$$\psi'_\alpha(x) = e^{ie\Lambda(x)}\psi_\alpha(x). \tag{3.50}$$

Analogously

$$\left(\frac{\partial}{\partial x_\mu} + ieA_\mu\right)\psi^*_\alpha = e^{ie\Lambda}\left(\frac{\partial}{\partial x_\mu} + ieA'_\mu\right)\psi^{*\prime}_\alpha, \tag{3.51}$$

where

$$\psi^{*\prime}_\alpha(x) = e^{-ie\Lambda(x)}\psi^*_\alpha(x). \tag{3.52}$$

Suppose that for $A_\mu = 0$ the Lagrangian of the system is invariant under the gauge transformations (3.35) with phase χ independent of x. If the potential A_μ enters the Lagrangian $\mathcal{L}'$ in the form (3.48), then from relationships (3.49) and (3.51) it is obvious that the Lagrangian of the system under consideration is invariant under transformations (3.46), (3.50), and (3.52) (these transformations are called *gauge transformations of the second kind*). Note that the requirement of invariance of the Lagrangian under the gauge transformations of the second kind uniquely determines the interaction Lagrangian of the electromagnetic field and the field ψ_α only when the derivatives of the potential A_μ do not appear in the Lagrangian $\mathcal{L}'$. This interaction is called the minimal electromagnetic interaction and e is a constant characterizing the interaction.

From the general equations (2.14) we obtain the following equations for the electromagnetic field:

$$\frac{\partial F_{\mu\nu}}{\partial x_\nu} = \frac{\partial\mathcal{L}}{\partial A_\mu}. \tag{3.53}$$

Comparing (3.53) and Maxwell's equations (2.31) we conclude that the electric current is

$$j^e_\mu = \frac{\partial\mathcal{L}}{\partial A_\mu}. \tag{3.54}$$

If one recognizes that the potential A_μ enters the Lagrangian $\mathcal{L}'$ in combination with the derivatives of the fields [see (3.48)], then it is not difficult to see that

$$j^e_\mu = -ie\left(\frac{\partial \mathcal{L}}{\partial \dfrac{\partial \psi_\alpha}{\partial x_\mu}}\psi_\alpha - \frac{\partial \mathcal{L}}{\partial \dfrac{\partial \psi^*_\alpha}{\partial x_\mu}}\psi^*_\alpha\right). \tag{3.55}$$

Thus, the current j^e_μ coincides with the current defined by relationship (3.40).

CHAPTER 3

FIELD QUANTIZATION

4. The Real Scalar (Pseudoscalar) Field

In the preceding chapter it was shown that the conserved quantities and the equations of the field can be obtained easily if the Lagrangian of the field is known. We shall now examine different types of fields. First the equations of the field will be postulated. Then we shall construct a Lagrangian which satisfies the requirements of Lorentz invariance under gauge transformations and which yield the postulated equations of the field. After that we shall obtain expressions for the conserved quantities from the general formulae of the preceding section.

We begin with the simplest case—real scalar and pseudoscalar fields. Denote a real function which transforms as either a scalar or a pseudoscalar [see (2.23) and (2.24)] by $\phi(x)$. We assume that the function $\phi(x)$ satisfies a linear Lorentz-invariant equation of the second order. The only such equation is the Klein–Gordon equation:

$$(\Box - \varkappa^2)\,\phi(x) = 0. \tag{4.1}$$

Here $\varkappa$ is a real positive parameter. In quantum theory, as we shall see below, $\varkappa$ is the rest mass of the field quantum. Note that the equation for the electromagnetic potential A_μ when $j_\mu = 0$ is a special case of equation (4.1) ($\varkappa = 0$). Correspondingly, quanta of the electromagnetic field (photons) are particles with rest mass equal to zero.

The Lagrangian of the field must be such that equation (4.1) coincides with

$$\frac{\partial\mathcal{L}}{\partial\phi} - \frac{\partial}{\partial x_\mu}\frac{\partial\mathcal{L}}{\partial\dfrac{\partial\phi}{\partial x_\mu}} = 0. \tag{4.2}$$

Obviously the term $\Box\,\phi$ in (4.1) can arise only from

$$\frac{\partial}{\partial x_\mu}\left(\frac{\partial\mathcal{L}}{\partial\dfrac{\partial\phi}{\partial x_\mu}}\right)$$

and the term $\varkappa^2\phi$ from $\partial\mathcal{L}/\partial\phi$. If these remarks are taken into consideration along with the requirement that the Lagrangian be scalar, it is easy to obtain the expression for $\mathcal{L}$ up to a common factor. It is not difficult to verify that the Lagrangian

$$\mathcal{L} = -\tfrac{1}{2}\left[\left(\frac{\partial\phi}{\partial x_\mu}\right)^2 + \varkappa^2\phi^2\right] \tag{4.3}$$

satisfies all the requirements. Actually,

$$\frac{\partial\mathcal{L}}{\partial\dfrac{\partial\phi}{\partial x_\mu}} = -\frac{\partial\phi}{\partial x_\mu}, \quad \frac{\partial\mathcal{L}}{\partial\phi} = -\varkappa^2\phi, \tag{4.4}$$

and from (4.2) we obtain the Klein–Gordon equation (4.1). Furthermore, it is easy to see that the Lagrangian (4.3) is a scalar. In fact, with the help of (2.23) [or (2.24) for the pseudoscalar field] we obtain

$$\begin{aligned}\mathcal{L}\left(\phi(x), \frac{\partial\phi(x)}{\partial x}\right) &= -\tfrac{1}{2}\left[\frac{\partial\phi'(x')}{\partial x'_\nu} a_{\nu\varrho} \frac{\partial\phi'(x')}{\partial x'_\sigma} a_{\sigma\varrho} + \varkappa^2\phi'(x')\,\phi'(x')\right] \\ &= -\tfrac{1}{2}\left[\left(\frac{\partial\phi'(x')}{\partial x'_\sigma}\right)^2 + \varkappa^2(\phi'(x'))^2\right] = \mathcal{L}\left(\phi'(x'), \frac{\partial\phi'(x')}{\partial x'}\right).\end{aligned} \tag{4.5}$$

Using (3.4) we find the following expression for the energy-momentum tensor of the scalar (pseudoscalar) field:

$$T_{\mu\nu} = -\frac{\partial\phi}{\partial x_\nu}\frac{\partial\phi}{\partial x_\mu} - \mathcal{L}\delta_{\mu\nu}. \tag{4.6}$$

From (3.3), (4.3), and (4.6) we find that the energy and momentum of the scalar (pseudoscalar) field are equal to

$$H = -iP_4 = \tfrac{1}{2}\int\left[\left(\frac{\partial\phi}{\partial x_0}\right)^2 + (\nabla\phi)^2 + \varkappa^2\phi^2\right] d\boldsymbol{x}, \tag{4.7}$$

$$P_k = -\int \frac{\partial\phi}{\partial x_0}\frac{\partial\phi}{\partial x_k}\, d\boldsymbol{x}. \tag{4.8}$$

The field we have been examining can be completely characterized by the real function $\phi(x)$ which satisfies the Klein–Gordon equation. Obviously the field can also be written in terms of the Fourier components of the function $\phi(x)$. For the transition to quantum theory it is more convenient to use this means to define the field. The Fourier expansion of the function $\phi(x)$ has the form

$$\phi(x) = \frac{1}{(2\pi)^{3/2}}\int \phi(\boldsymbol{q}, x_0)\, e^{i\boldsymbol{q}\boldsymbol{x}}\, d\boldsymbol{q}. \tag{4.9}$$

Obviously the coefficients of the expansion depend on time. Substituting (4.9) into the Klein–Gordon equation we find that the coefficients $\phi(\boldsymbol{q}, x_0)$ satisfy the equation

$$\frac{\partial^2\phi(\boldsymbol{q}, x_0)}{\partial x_0^2} + (\boldsymbol{q}^2 + \varkappa^2)\,\phi(\boldsymbol{q}, x_0) = 0. \tag{4.10}$$

The general solution of this equation can be written in the form

$$\phi(\boldsymbol{q}, x_0) = \phi^{(+)}(\boldsymbol{q})\, e^{-iq_0x_0} + \phi^{(-)}(\boldsymbol{q})\, e^{iq_0x_0}, \tag{4.11}$$

where

$$q_0 = +\sqrt{\boldsymbol{q}^2+\varkappa^2}.$$

Inserting (4.11) into (4.9) we find

$$\phi(x) = \phi^{(+)}(x)+\phi^{(-)}(x). \tag{4.12}$$

Here

$$\phi^{(+)}(x) = \frac{1}{(2\pi)^{3/2}}\int e^{iqx}\phi^{(+)}(\boldsymbol{q})\, d\boldsymbol{q}, \tag{4.12a}$$

$$\phi^{(-)}(x) = \frac{1}{(2\pi)^{3/2}}\int e^{-iqx}\phi^{(-)}(-\boldsymbol{q})\, d\boldsymbol{q}, \tag{4.12b}$$

where

$$q = (\boldsymbol{q}, iq_0); \quad qx = q_\mu x_\mu = \boldsymbol{qx} - q_0x_0.$$

Note that the expression (4.12b) is obtained by a change in the integration variables $(\boldsymbol{q} \to -\boldsymbol{q})$.

The function $\phi^{(+)}(x)$ is called the positive-frequency part of the function $\phi(x)$ (the integrand has the factor $e^{-iq_0x_0}$) and the function $\phi^{(-)}(x)$ is the negative-frequency part of $\phi(x)$ (the integrand has the factor $e^{-i(-q_0)x_0}$).

We now look at the consequences of the fact that the function $\phi(x)$ is real. It is obvious that

$$\phi^*(x) = \frac{1}{(2\pi)^{3/2}}\int (\phi^{(+)}(\boldsymbol{q}))^*\, e^{-iqx}\, d\boldsymbol{q} + \frac{1}{(2\pi)^{3/2}}\int (\phi^{(-)}(-\boldsymbol{q}))^*\, e^{iqx}\, d\boldsymbol{q}. \tag{4.13}$$

Equating (4.12) and (4.13) we obtain

$$\phi^{(-)}(-\boldsymbol{q}) = (\phi^{(+)}(\boldsymbol{q}))^*. \tag{4.14}$$

Thus, the real scalar (pseudoscalar) field is characterized by a complex function $\phi^{(+)}(\boldsymbol{q})$, in other words, by the Fourier components of the positive-frequency part of the function $\phi(x)$.

We express the energy and momentum of the field in terms of $\phi^{(+)}(\boldsymbol{q})$. Inserting the expansion (4.12) into (4.7) and integrating over $\boldsymbol{x}$, we obtain for the energy of the field

$$\begin{aligned} H = \tfrac{1}{2}\int [&\phi^{(+)}(\boldsymbol{q})\, \phi^{(+)}(-\boldsymbol{q})\, e^{-2iq_0x_0}(\boldsymbol{q}^2+\varkappa^2-q_0^2) + \phi^{(+)}(\boldsymbol{q})\, \phi^{(-)}(-\boldsymbol{q})\, (\boldsymbol{q}^2+\varkappa^2+q_0^2) \\ &+ \phi^{(-)}(-\boldsymbol{q})\, \phi^{(+)}(\boldsymbol{q})\, (\boldsymbol{q}^2+\varkappa^2+q_0^2) + \phi^{(-)}(-q)\, \phi^{(-)}(\boldsymbol{q})\, e^{2iq_0x_0}(\boldsymbol{q}^2+\varkappa^2-q_0^2)]\, d\boldsymbol{q}. \end{aligned} \tag{4.15}$$

If one recognizes that $q_0^2 = \boldsymbol{q}+\varkappa^2$, then from (4.15) we find

$$H = \int [\phi^{(+)}(\boldsymbol{q})\, \phi^{(-)}(-\boldsymbol{q}) + \phi^{(-)}(-\boldsymbol{q})\, \phi^{(+)}(\boldsymbol{q})]\, q_0^2\, d\boldsymbol{q}. \tag{4.16}$$

It is apparent that this integral does not depend on time (the terms containing $e^{\pm 2iq_0x_0}$ vanish).

Analogously, we obtain for the momentum of the field

$$P_k = +\int [\phi^{(+)}(\boldsymbol{q})\, \phi^{(+)}(-\boldsymbol{q})^{-2iq_0x_0}\, q_0q_k + \phi^{(+)}(\boldsymbol{q})\, \phi^{(-)}(-\boldsymbol{q}) + \phi^{(-)}(-\boldsymbol{q})\, \phi^{(+)}(\boldsymbol{q})\, q_0q_k + \phi^{(-)}(-\boldsymbol{q})\, \phi^{(-)}(\boldsymbol{q})\, e^{2iq_0x_0}]\, q_0q_k\, d\boldsymbol{q}. \qquad (4.17)$$

Obviously

$$\int \phi^{(\pm)}(\boldsymbol{q})\, \phi^{(\pm)}(-\boldsymbol{q})\, e^{\mp 2iq_0x_0} q_0q_k\, d\boldsymbol{q} = 0. \qquad (4.18)$$

Finally we find for the momentum of the field

$$P_k = \int [\phi^{(+)}(\boldsymbol{q})\, \phi^{(-)}(-\boldsymbol{q}) + \phi^{(-)}(-\boldsymbol{q})\, \phi^{(+)}(\boldsymbol{q})]\, q_0q_k\, d\boldsymbol{q}. \qquad (4.19)$$

Note that the first and second terms in the square brackets in (4.16) and (4.19) coincide in classical theory. Keeping in mind, however, the eventual transition to quantum theory, we shall leave the expressions for the energy and momentum in the forms (4.16) and (4.19). Furthermore, in place of the functions $\phi^{(+)}(\boldsymbol{q})$ and $\phi^{(-)}(-\boldsymbol{q})$ we introduce the functions $a(q)$ and $a(-q)$ defined by the relationships

$$\left.\begin{aligned} \phi^{(+)}(\boldsymbol{q}) &= \frac{1}{\sqrt{2q_0}}\, a(q), \\ \phi^{(-)}(-\boldsymbol{q}) &= \frac{1}{\sqrt{2q_0}}\, a(-q). \end{aligned}\right\} \qquad (4.20)$$

The Fourier expansion of the function $\phi(x)$ is written then in the form

$$\phi(x) = \frac{1}{(2\pi)^{3/2}} \int \frac{1}{\sqrt{2q_0}}\, a(q)\, e^{iqx}\, d\boldsymbol{q} + \frac{1}{(2\pi)^{3/2}} \int \frac{1}{\sqrt{2q_0}}\, a(-q)\, e^{-iqx}\, d\boldsymbol{q}. \qquad (4.21)$$

Inserting (4.20) into (4.16) and (4.19) we find for the energy and momentum of the field

$$H = \tfrac{1}{2} \int [a(q)\, a(-q) + a(-q)\, a(q)]\, q_0\, d\boldsymbol{q}, \qquad (4.16a)$$

$$P_k = \tfrac{1}{2} \int [a(q)\, a(-q) + a(-q)\, a(q)]\, q_k\, d\boldsymbol{q}, \qquad (4.19a)$$

where, as a result of (4.14),

$$a(-q) = a^*(q). \qquad (4.22)$$

Thus, the classical scalar (pseudoscalar) field is determined by the complex function $a(q)$, a function of the variable $\boldsymbol{q}$ which can take any value. The energy of the field [expression (4.16a)] can therefore take any positive value in the classical case.

We turn now to quantum field theory. Let us postulate that the quantized field is a set of free particles—the quanta of the field found in certain states. Specifically, we choose those states with definite monentum values. In accordance with general quantum theory postulates the states of a quantized field are to be described by wave functions which satisfy Schrödinger's equation and the energy, momentum, and other physical quantities are to be put in correspondence with operators whose eigenvalues are the possible values of the respective quantities. The energy operator must be chosen so that the functions describing free particles with definite momentum values are its eigenfunctions. As in ordinary quantum mechanics, the energy operator is constructed by replacing by operators those quantities which appear in the classical Hamiltonian. Thus, the classical quantities $a(q)$ and $a(-q)$ are to be replaced by operators.

If in expression (4.21) we replace $a(q)$ and $a(-q)$ by operators, we obtain† the field operator $\phi(x)$. It is clear that the operator $\phi(x)$ satisfies the Klein–Gordon equation.

The Hamiltonian must be a Hermitian operator. In order to ensure the Hermiticity of the Hamiltonian, it is necessary to replace the condition of reality of the classical field by the condition of Hermiticity of the operator $\phi(x)$, i.e. to require that

$$\phi^+(x) = \phi(x). \tag{4.23}$$

Then in place of (4.22) we obtain

$$a(-q) = a^+(q). \tag{4.24}$$

From (4.16) and (4.24) we find for the energy operator

$$H = \tfrac{1}{2} \int [a^+(q)\, a(q) + a(q)\, a^+(q)]\, q_0\, d\boldsymbol{q}. \tag{4.25}$$

It is obvious that this operator is Hermitian.

Furthermore, it is necessary to postulate commutation relations between the field operators. We assume the following commutation relations:

$$\left.\begin{aligned} [a(q), a(q')] &= a(q)\, a(q') - a(q')\, a(q) = 0, \\ [a(q), a^+(q')] &= a(q)\, a^+(q') - a^+(q')\, a(q) = \delta(q-q'). \end{aligned}\right\} \tag{4.26}$$

From this it also follows that

$$[a^+(q), a^+(q')] = 0. \tag{4.26a}$$

Using these commutation relations as a basis, we find the eigenfunctions of the Hamiltonian (4.25). We must make sure that the eigenfunctions of the Hamiltonian (4.25) describe free particles with definite momenta. Let us calculate the commutator $[a(q), H]$. With the help of the commutation relations (4.26) we find

$$[a(q), H] = \tfrac{1}{2} \int ([a(q), a^+(q')]\, a(q') + a(q')\, [a(q), a^+(q')])\, q_0'\, d\boldsymbol{q}' = q_0 a(q). \tag{4.27}$$

In obtaining (4.27) the relation

$$[\alpha, \beta\gamma] = \alpha\beta\gamma - \beta\gamma\alpha = (\alpha\beta - \beta\alpha)\, \gamma + \beta(\alpha\gamma - \gamma\alpha) = [\alpha, \beta]\, \gamma + \beta[\alpha, \gamma] \tag{4.28}$$

was used, where α, β, and γ are arbitrary operators. From (4.27) we also obtain by Hermitian conjugation

$$[a^+(q), H] = -q_0 a^+(q). \tag{4.29}$$

In connection with relations (4.27) and (4.29) we note the following. In calculating the commutator $[\phi, H]$, with the help of (4.27) and (4.29) one finds

$$[\phi(x), H] = \frac{1}{(2\pi)^{3/2}} \int e^{iqx} \frac{1}{\sqrt{2q_0}}\, q_0 a(q)\, d\boldsymbol{q} + \frac{1}{(2\pi)^{3/2}} \int e^{-iqx} \frac{1}{\sqrt{2q_0}} (-q_0)\, a^+(q)\, d\boldsymbol{q}. \tag{4.30}$$

† It is customary to use the same notation for operators in quantum theory as for the corresponding quantities in classical theory.

On the other hand, the right-hand side of relation (4.30) is equal to $i(\partial\phi/\partial x_0)$. Thus,

$$i\frac{\partial\phi(x)}{\partial x_0} = [\phi(x), H]. \tag{4.30a}$$

In this section we have been examining the free scalar (pseudoscalar) field. In the general case of several interacting fields relations (4.27) and (4.29) and, consequently, (4.30a) remain valid if H is understood to be the free Hamiltonian. Comparing (4.30a) and (1.15) we conclude that $\phi(x)$ is an operator in the interaction representation.

We now find the solution of the equation†

$$H\Phi_E = E\Phi_E, \tag{4.31}$$

where H is the Hamiltonian of the free scalar (pseudoscalar) field defined by expression (4.25). To do this we multiply the left-hand side by $a^+(q)$:

$$a^+(q)H\Phi_E = Ea^+(q)\Phi_E. \tag{4.32}$$

Obviously

$$a^+(q)H = [a^+(q), H]+Ha^+(q).$$

Using (4.29) we find

$$Ha^+(q)\Phi_E = (E+q_0)\, a^+(q)\Phi_E. \tag{4.33}$$

Thus, if Φ_E is an eigenvector of the Hamiltonian H which has the eigenvalue E, the state vector $a^+(q)\Phi_E$ is also an eigenvector of H and has the eigenvalue $E+q_0$.

Multiplying (4.31) by $a(q)$ we obtain

$$a(q)H\Phi_E = ([a(q), H]+Ha(q))\,\Phi_E = Ea(q)\,\Phi_E. \tag{4.34}$$

Using (4.27) we find

$$Ha(q)\,\Phi_E = (E-q_0)\, a(q)\,\Phi_E. \tag{4.35}$$

Hence, the state vector $a(q)\Phi_E$ is an eigenvector of the Hamiltonian H which has the eigenvalue $E-q_0$.

Furthermore, it is not difficult to see that all eigenvalues of the Hamiltonian (4.25) are positive. Actually, multiplying (4.31) on the left by Φ_E^+ we obtain

$$\frac{(\Phi_E^+ H\Phi_E)}{(\Phi_E^+\Phi_E)} = E. \tag{4.36}$$

Clearly, the denominator of this expression is positive. The numerator is also positive since

$$(\Phi_E^+ a^+(q)\, a(q)\,\Phi_E) = \big((a(q)\,\Phi_E)^+ a(q)\,\Phi_E\big) \geqslant 0. \tag{4.37}$$

Hence, the spectrum of eigenvalues of the Hamiltonian (4.25) is bounded from below, i.e. there exists a least eigenvalue. Denote the least eigenvalue of the Hamiltonian H by E_0 and

† We shall no longer use the concrete representation for the functions Φ_E which describe the state of the quantized field, but shall henceforth call them state vectors.

the state vector having the eigenvalue E_0 by Φ_0. Then it is obvious that

$$a(q)\Phi_0 = 0 \tag{4.38}$$

for all values of q (otherwise the vector $a(q)\Phi_0$ would have the eigenvalue $E_0 - q_0$).

From (4.25) and (4.36) we find for the least eigenvalue

$$E_0 = \frac{1}{2}\frac{1}{(\Phi_0^+\Phi_0)}\int [(\Phi_0^+ a^+(q)\, a(q)\Phi_0) + (\Phi_0^+\, a(q)\, a^+(q)\Phi_0)]\; q_0\, d\boldsymbol{q}. \tag{4.39}$$

As a result of (4.38) the first term in the square brackets is zero. It is further obvious that expression (4.39) can be written in the form

$$E_0 = \frac{1}{2(\Phi_0^+\Phi_0)}\int (\Phi_0^+ a(q)\, a^+(q')\Phi_0)\; \delta(\boldsymbol{q}' - \boldsymbol{q})\, q_0\, d\boldsymbol{q}'\, d\boldsymbol{q}. \tag{4.40}$$

Replacing here the operator $a(q)a^+(q')$ by $[a(q), a^+(q')] + a^+(q')a(q)$ and using (4.26) and (4.38) we obtain

$$E_0 = \tfrac{1}{2}\int \delta(\boldsymbol{q}' - \boldsymbol{q})\; \delta(\boldsymbol{q}' - \boldsymbol{q}) q_0\, d\boldsymbol{q}'\, d\boldsymbol{q}. \tag{4.41}$$

This integral diverges. We obtained therefore a meaningless result: the least eigenvalue of the Hamiltonian is infinite. In this connection, let us examine the choice of Hamiltonian. We assumed that the Hamiltonian in quantum theory can be obtained from the classical Hamiltonian by replacing the classical quantities $a(q)$ and $a^*(q)$ with the operators $a(q)$ and $a^+(q)$. Obviously this procedure is not unique.

In order to supplement the definition of the procedure for the transition from classical to quantum theory, we shall introduce the concept of the *normal product* of operators. Define in the following way the normal product operator N:

$$\left.\begin{aligned} N(a(q)\, a^+(q)) &= a^+(q)\, a(q), \\ N(a^+(q)\, a(q)) &= a^+(q)\, a(q). \end{aligned}\right\} \tag{4.42}$$

If the operator $a(q)$ is placed to the right of operator $a^+(q)$, then such an order of operators is called *normal*. Thus, the operator N acting on the product of two field operators arranges them in normal order. Below, the operator N will be defined in the general case of n factors.

We return now to the Hamiltonian in quantum theory. It is clear that the divergence of the least eigenvalue of the Hamiltonian (4.25) is related to the term $a(q)a^+(q)$. We shall assume that in obtaining the Hamiltonian in quantum theory from the classical Hamiltonian, besides replacing functions with operators, it is necessary to act on the product of field operators with the operator N. We arrive at the following Hamiltonian in quantum theory:

$$H = \int a^+(q)\, a(q)\, q_0\, d\boldsymbol{q}. \tag{4.43}$$

From (4.38) and (4.43) we find that

$$H\Phi_0 = 0. \tag{4.44}$$

Thus, the least eigenvalue of the Hamiltonian (4.43) is zero.

Let us assume that the operators which provide all other physical quantities are also obtained from corresponding classical quantities by replacing functions with operators arranged in normal order. From (4.19a) we obtain for the momentum operator

$$P_k = \int a^+(q)\, a(q) q_k\, d\boldsymbol{q}. \tag{4.45}$$

Obviously

$$P_k \Phi_0 = 0. \tag{4.46}$$

Thus, the vector Φ describes a state with energy and momentum equal to zero. Such a state is called a *vacuum*.

It is not difficult to see that the Hamiltonian (4.43) satisfies relations (4.27) and (4.29). Consequently, from equation (4.33) we find that the vector $a^+(q_1)\Phi_0$ is an eigenvector of the Hamiltonian which has the eigenvalue q_{10}. We shall show that this vector is also an eigenvector $\boldsymbol{P}$ of the momentum operator. With the help of (4.26) and (4.45) we find

$$[\boldsymbol{P}, a^+(q)] = \int a^+(q')\, [a(q'), a^+(q)] \boldsymbol{q}'\, d\boldsymbol{q}' = \boldsymbol{q} a^+(q). \tag{4.47}$$

We further obtain

$$\boldsymbol{P} a^+(q_1)\Phi_0 = ([\boldsymbol{P}, a^+(q_1)] + a^+(q_1)\boldsymbol{P})\Phi_0 = \boldsymbol{q}_1 a^+(q_1)\, \Phi_0. \tag{4.48}$$

Thus, the state vector $a^+(q_1)\Phi_0$ describes a spinless particle (field quantum) with energy q_{10} and momentum $\boldsymbol{q}_1$. The mass of this particle is equal to $\sqrt{q_{10}^2 - \boldsymbol{q}_1^2} = \varkappa$.

If we operate on the state vector $a^+(q_1)\Phi_0$ with the operator $a^+(q_2)$, we obtain an eigenvector of the Hamiltonian (4.43) with eigenvalue $q_{10}+q_{20}$. The vector $a^+(q_2)a^+(q_1)\Phi_0$ is also an eigenvector of the momentum operator with eigenvalue $\boldsymbol{q}_1+\boldsymbol{q}_2$. Actually, using (4.47) and (4.48) we obtain

$$\boldsymbol{P} a^+(q_2)\, a^+(q_1)\, \Phi_0 = ([\boldsymbol{P}, a^+(q_2)] + a^+(q_2)\, \boldsymbol{P})\, a^+(q_1)\, \Phi_0 = (\boldsymbol{q}_1 + \boldsymbol{q}_2)\, a^+(q_2)\, a^+(q_1)\Phi_0. \tag{4.49}$$

The state vector $a^+(q_2)a^+(q_1)\Phi_0$ describes therefore two non-interacting particles with four-momenta q_1 and q_2 and the same mass $\varkappa$. If we act on the vector $a^+(q_2)a^+(q_1)\Phi_0$ with the operator $a^+(q_3)$, we obtain a state vector which describes three non-interacting particles with momenta q_1, q_2, q_3, etc.

Thus, by acting on the vacuum state vector Φ_0 with the operators a^+, we obtain a set of eigenvectors of the Hamiltonian, which is a set of state vectors of identical non-interacting particles with mass $\varkappa$, spin zero, and definite values for their four-momentum. The operators $a^+(q)$ and $a(q)$ are called respectively *creation* and *annihilation* operators of a particle with four-momentum q. In the general case the eigenvector of the Hamiltonian (4.43) can be written in the form

$$\Phi_{n(q_n)\ldots n(q_1)} = \underbrace{a^+(q_n)\ldots a^+(q_n)}_{n(q_n)} \ldots \underbrace{a^+(q_1)\ldots a^+(q_1)}_{n(q_1)}\Phi_0. \tag{4.50}$$

Here $n(q_1)$ is the number of operators $a^+(q_1)$; $n(q_2)$ is the number of operators $a^+(q_2)$, and so forth. The vector (4.50) describes a state in which $n(q_1)$, $n(q_2)$, and so forth of the identical particles have the respective four-momenta q_1, q_2, and so forth. The numbers $n(q_1)$, $n(q_2)$ are called *occupation numbers*. It is obvious that the state vector is uniquely given by the occupation numbers. The state vector (4.50) has the following values for the total energy and momentum:

$$\left.\begin{aligned} E_{n(q_n)\dots n(q_1)} &= \sum_{i=1}^{n} n(q_i)\, q_{i0}, \\ \boldsymbol{P}_{n(q_n)\dots n(q_1)} &= \sum_{i=1}^{n} n(q_i)\, \boldsymbol{q}_i. \end{aligned}\right\} \tag{4.51}$$

All of the above is true for scalar as well as for pseudoscalar fields. Now let us examine how the quantum theory of these fields and the corresponding particles differs.

We shall carry out an inversion of the reference system. Let $\phi'(x')$ be the function which describes the field in the new reference system. Then

$$\phi'(x') = \eta_P \phi(x), \tag{4.52}$$

where

$$\boldsymbol{x}' = -\boldsymbol{x}, \quad x_0' = x_0, \tag{4.53}$$

and η_P is a phase factor. If we again invert the reference system, we return to the original system. Here we obtain

$$\phi(x) = \eta_P \phi'(x'). \tag{4.54}$$

From (4.52) and (4.54) we conclude that

$$\eta_P^2 = 1, \tag{4.55}$$

i.e. $\eta_P = \pm 1$. The quantity η_P is called the *intrinsic parity*. For the scalar field $\eta_P = 1$, for the pseudoscalar field $\eta_P = -1$.

Let us now examine the transformation of the operators of quantum field theory under inversion of the reference system. The field operator in the primed system at point x' is equal to

$$\phi'(x') = U_P \phi(x') U_P^{-1}, \tag{4.56}$$

where U_P is a unitary operator ($U_P^+ U_P = 1$) which acts on the field operators. In accordance with (4.52) we obtain the transformation of the field operators under inversion of the reference system

$$U_P \phi(x') U_P^{-1} = \eta_P \phi(x). \tag{4.57}$$

Inserting the expansion (4.21) into (4.57) we find

$$\begin{aligned} U_P a(q) U_P^{-1} &= \eta_P a(q'), \\ U_P a^+(q) U_P^{-1} &= \eta_P a^+(q'), \end{aligned} \tag{4.58}$$

where

$$\boldsymbol{q}' = -\boldsymbol{q}, \quad q_0' = q_0. \tag{4.59}$$

Let us act on the state vector $\Phi_q = a^+(q)\Phi_0$, which describes a particle with momentum q, with the operator U_P. We obtain

$$U_P \Phi_q = U_P a^+(q) U_P^{-1} U_P \Phi_0. \tag{4.60}$$

It is natural to require that

$$U_P \Phi_0 = \Phi_0. \tag{4.61}$$

Then we find

$$U_P \Phi_q = \eta_P \Phi_{q'}. \tag{4.62}$$

Thus, if we act on the state vector which describes a particle with four-momentum q by the operator U_P, we obtain a state vector of the same particle with four-momentum q', multiplied by the intrinsic parity η_P. If $\eta_P = 1$, the particle is called scalar. If $\eta_P = -1$, the particle is called pseudoscalar. The quantity η_P characterizes the intrinsic properties of the particles and is determined experimentally. It is obvious that

$$U_P \Phi_{n(q_n)\dots n(q_1)} = (\eta_P)^{n(q_1)+\dots+n(q_n)} \Phi_{n(q_n')\dots n(q_1')}, \tag{4.63}$$

where

$$q_i' = -q_i; \quad q_{i0}' = q_{i0}.$$

5. The Complex Scalar (Pseudoscalar) Field

In this section the complex scalar (pseudoscalar) field will be examined. We begin with the classical case. Let us assume that the complex field function $\phi(x)$ satisfies the Klein–Gordon equation

$$(\Box - \varkappa^2)\, \phi(x) = 0. \tag{5.1}$$

First we shall construct a Lagrangian which gives the postulated equation of the field (5.1) and satisfies the requirements of relativistic invariance and invariance under gauge transformations. Then, using the general formulae of § 3, we shall find the conserved quantities. Finally, we shall make the transition to quantum theory and show that the quanta of the complex field are charged particles with charge e and $-e$.

It is obvious that the complex function $\phi(x)$ can always be put in the form

$$\phi(x) = \phi_1(x) + i\phi_2(x),$$

where

$$\phi_1 = \frac{\phi + \phi^*}{2} \quad \text{and} \quad \phi_2 = \frac{\phi - \phi^*}{2i} \tag{5.2}$$

are the real and imaginary parts of the function $\phi(x)$.

Thus, this field is described by two independent functions. We shall describe the field by the functions $\phi(x)$ and $\phi^*(x)$. From (5.1) it follows that

$$(\Box - \varkappa^2)\, \phi^*(x) = 0. \tag{5.1a}$$

On the other hand, the equations of the complex field are [see (2.14) and (2.15)]

$$\frac{\partial \mathscr{L}}{\partial \phi} - \frac{\partial}{\partial x_\mu} \frac{\partial \mathscr{L}}{\partial \frac{\partial \phi}{\partial x_\mu}} = 0, \quad \frac{\partial \mathscr{L}}{\partial \phi^*} - \frac{\partial}{\partial x_\mu} \frac{\partial \mathscr{L}}{\partial \frac{\partial \phi^*}{\partial x_\mu}} = 0. \tag{5.3}$$

The Lagrangian depends on the functions ϕ, ϕ^*, and their first derivatives. It is clear that the terms $\Box\phi$ and $\Box\phi^*$ in (5.1) come from

$$\frac{\partial}{\partial x_\mu}\frac{\partial\mathcal{L}}{\partial\frac{\partial\phi}{\partial x_\mu}} \quad\text{and}\quad \frac{\partial}{\partial x_\mu}\frac{\partial\mathcal{L}}{\partial\frac{\partial\phi^*}{\partial x_\mu}}.$$

Taking into account the requirement of Lorentz invariance, we conclude that the Lagrangian must be a linear combination of the quantities

$$\frac{\partial\phi}{\partial x_\varrho}\frac{\partial\phi}{\partial x_\varrho}, \quad \frac{\partial\phi^*}{\partial x_\varrho}\frac{\partial\phi^*}{\partial x_\varrho}, \quad \frac{\partial\phi^*}{\partial x_\varrho}\frac{\partial\phi}{\partial x_\varrho}, \quad \phi\phi, \quad \phi^*\phi^*, \quad \phi^*\phi. \tag{5.4}$$

Only the third and sixth terms in (5.4) satisfy the requirements of gauge invariance. If one keeps these remarks in mind, it is not difficult to obtain an expression for $\mathcal{L}$ within a common factor. We take the following expression for the Lagrangian of the complex scalar (pseudoscalar) field:

$$\mathcal{L} = -\frac{\partial\phi^*}{\partial x_\varrho}\frac{\partial\phi}{\partial x_\varrho} - \varkappa^2\phi^*\phi. \tag{5.5}$$

This Lagrangian satisfies all invariance requirements.

From (5.5) it also follows that the functions ϕ and ϕ^* obey the Klein–Gordon equation. Actually,

$$\left.\begin{aligned} &\frac{\partial\mathcal{L}}{\partial\phi} = -\varkappa^2\phi^*, && \frac{\partial\mathcal{L}}{\partial\phi^*} = -\varkappa^2\phi, \\ &\frac{\partial\mathcal{L}}{\partial\frac{\partial\phi}{\partial x_\mu}} = -\frac{\partial\phi^*}{\partial x_\mu}, && \frac{\partial\mathcal{L}}{\partial\frac{\partial\phi^*}{\partial x_\mu}} = -\frac{\partial\phi}{\partial x_\mu}. \end{aligned}\right\} \tag{5.6}$$

Inserting (5.6) into (5.3) we obtain equation (5.1).

Let us calculate the energy-momentum tensor [see (3.4)]. We have

$$T_{\mu\nu} = \frac{\partial\mathcal{L}}{\partial\frac{\partial\phi}{\partial x_\nu}}\frac{\partial\phi}{\partial x_\mu} + \frac{\partial\mathcal{L}}{\partial\frac{\partial\phi^*}{\partial x_\nu}}\frac{\partial\phi^*}{\partial x_\mu} - \mathcal{L}\delta_{\mu\nu} = -\frac{\partial\phi^*}{\partial x_\nu}\frac{\partial\phi}{\partial x_\mu} - \frac{\partial\phi}{\partial x_\nu}\frac{\partial\phi^*}{\partial x_\mu} - \mathcal{L}\delta_{\mu\nu}. \tag{5.7}$$

From (5.7) we find for the energy and momentum of the field

$$\left.\begin{aligned} H &= \int\left(\nabla\phi^*\,\nabla\phi + \varkappa^2\phi^*\phi + \frac{\partial\phi^*}{\partial x_0}\frac{\partial\phi}{\partial x_0}\right)dx, \\ P_k &= -\int\left(\frac{\partial\phi^*}{\partial x_0}\frac{\partial\phi}{\partial x_k} + \frac{\partial\phi}{\partial x_0}\frac{\partial\phi^*}{\partial x_k}\right)dx. \end{aligned}\right\} \tag{5.8}$$

Furthermore, from (3.40) we obtain for the current four-vector

$$j_\mu = -ie\left(\frac{\partial\mathcal{L}}{\partial\frac{\partial\phi}{\partial x_\mu}}\phi - \frac{\partial\mathcal{L}}{\partial\frac{\partial\phi^*}{\partial x_\mu}}\phi^*\right) = ie\left(\frac{\partial\phi^*}{\partial x_\mu}\phi - \frac{\partial\phi}{\partial x_\mu}\phi^*\right). \tag{5.9}$$

The charge of the field is given by the expression

$$Q = -i \int j_4(x)\, dx. \tag{5.10}$$

Expand the function $\phi(x)$ in a Fourier integral. Using the equation of the field (5.1) we obtain

$$\phi(x) = \phi^{(+)}(x) + \phi^{(-)}(x), \tag{5.11}$$

where

$$\left.\begin{aligned} \phi^{(+)}(x) &= \frac{1}{(2\pi)^{3/2}} \int \frac{1}{\sqrt{2q_0}}\, a(q)\, e^{iqx}\, d\boldsymbol{q}, \\ \phi^{(-)}(x) &= \frac{1}{(2\pi)^{3/2}} \int \frac{1}{\sqrt{2q_0}}\, a(-q)\, e^{-iqx}\, d\boldsymbol{q}, \\ q_0 &= \sqrt{\varkappa^2 + \boldsymbol{q}^2}. \end{aligned}\right\} \tag{5.12}$$

The essential difference from the case of the real field discussed earlier is the fact that the functions $a(q)$ and $a(-q)$ in this case are independent. Consequently, the complex scalar (pseudoscalar) field can be described by two complex functions of the variable $\boldsymbol{q}$.

Inserting (5.11) and (5.12) into (5.8) and (5.10) we obtain the following expressions for the energy-momentum four-vector and the charge of the field:

$$P_\mu = \int [a^*(q)\, a(q) + a^*(-q)\, a(-q)]\, q_\mu\, d\boldsymbol{q}, \tag{5.13}$$

$$Q = e \int [a^*(q)\, a(q) - a^*(-q)\, a(-q)]\, d\boldsymbol{q}. \tag{5.14}$$

From (5.12) it follows that the energy of the field can assume only positive values. The charge of the field Q can be either positive or negative.

We turn now to quantum theory. The quantities $a(q)$ and $a(-q)$ in classical theory are to be replaced by operators as we did for the real (Hermitian) field. The operation of complex conjugation of functions is to be replaced by the Hermitian conjugation of operators. Clearly, the energy, momentum, and charge operators obtained by such a method will be Hermitian.

The field operator $\phi(x)$, which satisfies the Klein–Gordon equation, is an operator in the interaction representation. All the operators in the interaction representation obey relationships (1.15). Thus, for $\phi(x)$ we obtain

$$i \frac{\partial \phi(x)}{\partial x_0} = [\phi(x), H]. \tag{5.15}$$

[H_0 in relationships (1.15) coincides with H for the free field we are examining.] Inserting the expansion (5.12) into (5.15) [$a(q)$ and $a(-q)$ are operators] we obtain

$$\frac{1}{(2\pi)^{3/2}} \int [(q_0 a(q) - [a(q), H])\, e^{iqx} + (-q_0 a(-q) - [a(-q), H])\, e^{-iqx}] \frac{d\boldsymbol{q}}{\sqrt{2q_0}} = 0. \tag{5.16}$$

Hence

$$[a(q), H] = q_0 a(q), \quad [a(-q), H] = -q_0 a(-q). \tag{5.17}$$

By Hermitian conjugation we obtain from these relationships

$$[a^+(q), H] = -q_0 a^+(q), \quad [a^+(-q), H] = q_0 a^+(-q). \tag{5.17a}$$

With the help of (5.17) it is not difficult to verify (see preceding paragraph) that the operators $a^+(q)$ and $a(-q)$ [$a(q)$ and $a^+(-q)$], when acting on the eigenvector of the Hamiltonian H which has energy E, give an eigenvector of the operator H with energy $E+q_0$ ($E-q_0$). Thus, the Hermitian field $\phi(x)$ has two creation operators [$a^+(q)$ and $a(-q)$] and two annihilation operators [$a(q)$ and $a^+(-q)$]. Denote

$$a(-q) = b^+(q). \tag{5.18}$$

Hence

$$a^+(-q) = b(q).$$

Using expressions (5.12) and (5.14) and writing the field operators in normal order (the annihilation operator to the right of the creation operator), we obtain the following expressions for the energy, momentum, and charge operators:

$$H = \int [a^+(q)\, a(q) + b^+(q)\, b(q)]\, q_0\, d\boldsymbol{q}, \tag{5.19}$$

$$P_k = \int [a^+(q)\, a(q) + b^+(q)\, b(q)]\, q_k\, d\boldsymbol{q}, \tag{5.20}$$

$$Q = e \int [a^+(q)\, a(q) - b^+(q)\, b(q)]\, d\boldsymbol{q}. \tag{5.21}$$

It is obvious that all eigenvalues of the Hamiltonian (5.19) are positive. If the eigenvector of the Hamiltonian H with the least eigenvalue is denoted by Φ_0, it is clear that

$$a(q)\Phi_0 = 0, \quad b(q)\Phi_0 = 0. \tag{5.22}$$

From (5.19)–(5.22) we find

$$H\Phi_0 = 0, \quad \boldsymbol{P}\Phi_0 = 0, \quad Q\Phi_0 = 0. \tag{5.23}$$

Thus, the vector Φ_0 describes the state with energy, momentum, and charge equal to zero (the vacuum state).

For the field operators under consideration we postulate the following commutation relations:

$$\left.\begin{aligned} [a(q), a(q')] &= 0, \\ [b(q), b(q')] &= 0, \\ [a(q), a^+(q')] &= \delta(\boldsymbol{q}-\boldsymbol{q}'), \\ [b(q), b^+(q')] &= \delta(\boldsymbol{q}-\boldsymbol{q}'), \\ [a(q), b(q')] &= 0, \\ [a(q), b^+(q')] &= 0. \end{aligned}\right\} \tag{5.24}$$

From these relations it follows that

$$\left.\begin{aligned} [a^+(q), a^+(q')] &= 0, \\ [b^+(q), b^+(q')] &= 0, \\ [a^+(q), b^+(q')] &= 0, \\ [a^+(q), b(q')] &= 0. \end{aligned}\right\} \tag{5.24a}$$

Using (5.24) and (5.19) it is not difficult to verify that relations (5.17) are true. Actually, with the help of relations (4.28) we obtain

$$[a(q), H] = \int [a(q), a^+(q')]a(q')q_0' \, d\boldsymbol{q}' = q_0 a(q),$$
$$[b(q), H] = \int [b(q), b^+(q')]b(q')q_0' \, d\boldsymbol{q} = q_0 b(q).$$

Analogously we find

$$[a(q), \boldsymbol{P}] = \boldsymbol{q}a(q), \quad [b(q), \boldsymbol{P}] = \boldsymbol{q}b(q). \tag{5.25}$$

It follows from this ($\boldsymbol{P}$ is a Hermitian operator) that

$$[\boldsymbol{P}, a^+(q)] = \boldsymbol{q}a^+(q), \quad [\boldsymbol{P}, b^+(q)] = \boldsymbol{q}b^+(q). \tag{5.25a}$$

We turn now to constructing a set of eigenvectors of the Hamiltonian (5.19). Consider the state vector $a^+(q)\Phi_0$. Using (5.17) and (5.23) we obtain

$$Ha^+(q)\Phi_0 = ([H, a^+(q)]+a^+(q)H)\Phi_0 = q_0 a^+(q)\Phi_0, \tag{5.26}$$
$$\boldsymbol{P}a^+(q)\Phi_0 = ([\boldsymbol{P}, a^+(q)]+a^+(q)\boldsymbol{P})\Phi_0 = \boldsymbol{q}a^+(q)\Phi_0. \tag{5.27}$$

The vector $a^+(q)\Phi_0$ describes consequently a particle with four-momentum q and mass $\sqrt{q_0^2-\boldsymbol{q}^2} = \varkappa$. Analogously we find

$$\left.\begin{aligned} Hb^+(q)\Phi_0 &= q_0 b^+(q)\Phi_0, \\ \boldsymbol{P}b^+(q)\Phi_0 &= \boldsymbol{q}b^+(b)\Phi_0. \end{aligned}\right\} \tag{5.28}$$

Hence, $b^+(q)\Phi$ is the state vector of a particle with four-momentum q and mass $\sqrt{q_0^2-\boldsymbol{q}^2}=\varkappa$.

Thus, $a^+(q)$ and $b^+(q)$ are the creation operators of particles with the same mass $\varkappa$. Let us see how these particles differ. We shall calculate the commutators $[Q, a^+(q)]$ and $[Q, b^+(q)]$. Using the commutation relations (5.24) and expression (5.21) we obtain

$$[Q, a^+(q)] = ea^+(q), \quad [Q, b^+(q)] = -eb^+(q). \tag{5.29}$$

We further have

$$\left.\begin{aligned} Qa^+(q)\Phi_0 &= ([Q, a^+(q)]+a^+(q)Q)\Phi_0 = ea^+(q)\Phi_0, \\ Qb^+(q)\Phi_0 &= ([Q, b^+(q)]+b^+(q)Q)\Phi_0 = -eb^+(q)\Phi_0. \end{aligned}\right\} \tag{5.30}$$

The state vector $a^+(q)\Phi_0$ describes therefore a charged particle with charge e and the vector $b^+(q)\Phi_0$ describes a particle with charge $-e$. Thus, if the charge of the particles —quanta of the field—is different from zero, then along with the particles with charge e and mass $\varkappa$ there must exist particles with the same mass $\varkappa$ and charge $-e$ (antiparticles). This important general conclusion is confirmed by experiment.

Let us act on the vector $a^+(q)\Phi_0$ with the operator $b^+(q')$. It is obvious that

$$\left.\begin{aligned} P_\mu b^+(q')a^+(q)\Phi_0 &= ([P_\mu, b^+(q')]+b^+(q')P_\mu)\, a^+(q)\Phi_0 \\ &= (q_\mu'+q_\mu)b^+(q')a^+(q)\Phi_0, \\ Qb^+(q')a^+(q)\Phi_0 &= [(-e)+(e)]b^+(q')a^+(q)\Phi_0. \end{aligned}\right\} \tag{5.31}$$

The state vector $b^+(q')a^+(q)\Phi_0$ describes therefore a particle with four-momentum q_μ and an antiparticle with four-momentum q_μ'.

Thus, if the vacuum state vector Φ_0 is acted upon with the creation operators a^+ and b^+, we obtain a set of eigenvectors of the Hamiltonian (5.19) which describes particles and antiparticles with definite momenta. In the general case we have

$$\begin{aligned}\Phi_{\bar{n}(q'_m)\ldots\bar{n}(q'_1)\,n(q_n)\ldots n(q_1)} = & \underbrace{b^+(q'_m)\ldots b^+(q'_m)}_{\bar{n}(q'_m)}\ldots \\ & \ldots\underbrace{b^+(q'_1)\ldots b^+(q'_1)}_{\bar{n}(q'_1)}\underbrace{a^+(q_n)\ldots a^+(q_n)}_{n(q_n)}\ldots\underbrace{a^+(q_1)\ldots a^+(q_1)}_{n(q_1)}\Phi_0.\end{aligned} \tag{5.32}$$

The occupation numbers of the particles $n(q_1), \ldots, n(q_n)$ and the occupation numbers of the antiparticles $n(q'_1), \ldots, n(q'_m)$ completely specify the state of the system. It is obvious that the vector (5.32) describes a state with the following values for the energy-momentum vector and the charge:

$$\left.\begin{aligned}P_\mu &= \sum_{i=1}^{n}(q_i)_\mu\, n(q_i) + \sum_{k=1}^{m}(q'_k)_\mu\, \bar{n}(q'_k), \\ Q &= e\sum_{i=1}^{n} n(q_i) + (-e)\sum_{k=1}^{m}\bar{n}(q'_k).\end{aligned}\right\} \tag{5.33}$$

6. The Complex Spinor Field

We shall now examine the field described by a complex spinor function $\psi_\sigma(x)$ (the index σ assumes four values) which obeys the Dirac equation:

$$\gamma_\mu \frac{\partial \psi}{\partial x_\mu} + m\psi = 0, \tag{6.1}$$

$$\frac{\partial \bar{\psi}}{\partial x_\mu}\gamma_\mu - m\bar{\psi} = 0. \tag{6.1a}$$

Here γ_μ are the Hermitian 4×4 matrices ($\gamma_\mu\gamma_\nu + \gamma_\nu\gamma_\mu = 2\delta_{\mu\nu}$); $\bar{\psi} = \psi^+\gamma_4$, and m is a positive parameter (in quantum theory it is the mass of the field quantum).

We choose as independent dynamic variables the functions ψ and $\bar{\psi}$. The equations of the field are [see (2.14)]

$$\frac{\partial \mathcal{L}}{\partial \psi_\sigma} - \frac{\partial}{\partial x_\mu}\frac{\partial \mathcal{L}}{\partial \dfrac{\partial \psi_\sigma}{\partial x_\mu}} = 0, \qquad \frac{\partial \mathcal{L}}{\partial \bar{\psi}_\sigma} - \frac{\partial}{\partial x_\mu}\frac{\partial \mathcal{L}}{\partial \dfrac{\partial \bar{\psi}_\sigma}{\partial x_\mu}} = 0. \tag{6.2}$$

We shall show that the Lagrangian

$$\mathcal{L} = -\tfrac{1}{2}\left[\left(\bar{\psi}\gamma_\varrho \frac{\partial \psi}{\partial x_\varrho} + m\bar{\psi}\psi\right) - \left(\frac{\partial \bar{\psi}}{\partial x_\varrho}\gamma_\varrho\psi - m\bar{\psi}\psi\right)\right] \tag{6.3}$$

satisfies the requirements of Lorentz invariance and invariance under gauge transformations and gives when substituted in (6.2) the equations of the field (6.1). Using (2.26) and

(2.27) we obtain

$$\left.\begin{aligned}\bar{\psi}(x)\gamma_\varrho \frac{\partial\psi(x)}{\partial x_\varrho} &= \bar{\psi}'(x')\, L\gamma_\varrho L^{-1}a_{\nu\varrho}\frac{\partial\psi'(x')}{\partial x'_\nu} = \bar{\psi}'(x')\,\gamma_\nu \frac{\partial\psi'(x')}{\partial x'_\nu},\\ \bar{\psi}(x)\,\psi(x) &= \bar{\psi}'(x')\,LL^{-1}\psi'(x') = \bar{\psi}'(x')\,\psi(x'),\\ \frac{\partial\bar{\psi}(x)}{\partial x_\varrho}\gamma_\varrho\psi(x) &= \frac{\partial\bar{\psi}'(x')}{\partial x'_\nu}L\gamma_\varrho L^{-1}a_{\nu\varrho}\psi'(x') = \frac{\partial\bar{\psi}'(x')}{\partial x'_\nu}\gamma_\nu\psi'(x').\end{aligned}\right\} \tag{6.4}$$

From (6.3) and (6.4) it follows that

$$\mathcal{L}(\psi(x), \bar{\psi}(x), \ldots) = \mathcal{L}(\psi'(x'), \bar{\psi}'(x'), \ldots), \tag{6.5}$$

i.e. the Lagrangian (6.3) is scalar. It is further obvious that the Lagrangian (6.3) is invariant under the gauge transformations

$$\left.\begin{aligned}\psi'(x) &= \psi(x)\, e^{ie\chi},\\ \bar{\psi}'(x) &= \bar{\psi}(x)\, e^{-ie\chi}\end{aligned}\right\} \tag{6.6}$$

(χ does not depend on x). Finally, from (6.3) we find

$$\left.\begin{aligned}\frac{\partial\mathcal{L}}{\partial\bar{\psi}_\sigma} &= -\tfrac{1}{2}\left(\gamma_\mu\frac{\partial\psi}{\partial x_\mu}\right)_\sigma - m\psi_\sigma,\\ \frac{\partial\mathcal{L}}{\partial\dfrac{\partial\bar{\psi}_\sigma}{\partial x_\mu}} &= \tfrac{1}{2}\,(\gamma_\mu\psi)_\sigma.\end{aligned}\right\} \tag{6.7}$$

Inserting (6.7) into (6.2) we obtain the Dirac equation for the functions $\psi(x)$:

$$\left(\gamma_\mu\frac{\partial\psi}{\partial x_\mu}\right)_\sigma + m\psi_\sigma = 0.$$

Furthermore,

$$\left.\begin{aligned}\frac{\partial\mathcal{L}}{\partial\psi_\sigma} &= -m\bar{\psi}_\sigma + \tfrac{1}{2}\left(\frac{\partial\bar{\psi}}{\partial x_\mu}\gamma_\mu\right)_\sigma,\\ \frac{\partial\mathcal{L}}{\partial\dfrac{\partial\psi_\sigma}{\partial x_\mu}} &= -\tfrac{1}{2}\,(\bar{\psi}\gamma_\mu)_\sigma,\end{aligned}\right\} \tag{6.8}$$

and from (6.2) we obtain the Dirac equation for the conjugate spinor $\bar{\psi}(x)$:

$$\left(\frac{\partial\bar{\psi}}{\partial x_\mu}\gamma_\mu\right)_\sigma - m\bar{\psi}_\sigma = 0.$$

From the general expression (3.4) it follows that the energy-momentum tensor of the spinor field is

$$T_{\mu\nu} = \frac{\partial\mathcal{L}}{\partial\dfrac{\partial\psi_\sigma}{\partial x_\nu}}\frac{\partial\psi_\sigma}{\partial x_\mu} + \frac{\partial\mathcal{L}}{\partial\dfrac{\bar{\psi}_\sigma}{\partial x_\nu}}\frac{\partial\bar{\psi}_\sigma}{\partial x_\mu} - \mathcal{L}\delta_{\mu\nu}. \tag{6.9}$$

Inserting (6.7) and (6.8) here and recognizing that the value of the Lagrangian (6.3) is zero for ψ and $\bar{\psi}$ satisfying equations (6.1) and (6.1a), we obtain

$$T_{\mu\nu} = -\tfrac{1}{2}\,\bar{\psi}\gamma_\nu \frac{\partial \psi}{\partial x_\mu} + \tfrac{1}{2}\,\frac{\partial \bar{\psi}}{\partial x_\mu}\gamma_\nu \psi. \tag{6.10}$$

In accordance with (3.40) the current vector of the complex spinor field is given by the expression

$$j_\mu = -ie\left(\frac{\partial \mathcal{L}}{\partial \dfrac{\partial \psi_\sigma}{\partial x_\mu}}\psi_\sigma - \frac{\partial \mathcal{L}}{\partial \dfrac{\partial \bar{\psi}_\sigma}{\partial x_\mu}}\bar{\psi}_\sigma\right) = ie\bar{\psi}\gamma_\mu\psi. \tag{6.11}$$

From this we find for the charge of the field

$$Q = e\int \bar{\psi}\gamma_4\psi\, dx. \tag{6.12}$$

Using (6.10) we find that the energy-momentum four-vector of the field is

$$\begin{aligned} P_\mu = i\int T_{\mu 4}\, dx &= -\frac{i}{2}\int \bar{\psi}\gamma_4 \frac{\partial \psi}{\partial x_\mu}\, dx + \frac{i}{2}\int \frac{\partial \bar{\psi}}{\partial x_\mu}\gamma_4\psi\, dx \\ &= -i\int \bar{\psi}\gamma_4 \frac{\partial \psi}{\partial x_\mu}\, dx + \frac{i}{2}\int \frac{\partial}{\partial x_\mu}(\bar{\psi}\gamma_4\psi)\, dx. \end{aligned} \tag{6.13}$$

It is not difficult to see that the second term on the right-hand side of (6.13) is zero. Actually, as a result of the conservation of the total charge

$$\frac{\partial}{\partial x_\mu}\int \bar{\psi}\gamma_4\psi\, dx = 0. \tag{6.14}$$

It is further obvious that

$$\int \frac{\partial}{\partial x_k}(\bar{\psi}\gamma_4\psi)\, dx = 0$$

(the absence of fields at infinity). Thus, the energy-momentum vector of the complex spinor field is

$$P_\mu = -i\int \bar{\psi}\gamma_4 \frac{\partial \psi}{\partial x_\mu}\, dx. \tag{6.15}$$

We shall now expand the function $\psi(x)$ in states with definite momentum. For a fixed momentum value the Dirac equation has, as is known, four independent solutions—two solutions with positive energy and two with negative energy. These solutions are discussed in detail in the Appendix. We write the solutions with positive energy in the form

$$\frac{1}{(2\pi)^{3/2}}\, u_+^r(\boldsymbol{p})\, e^{i\boldsymbol{p}\boldsymbol{x} - ip_0x_0}, \tag{6.16}$$

where $p_0 = +\sqrt{\boldsymbol{p}^2 + m^2}$ and the spinor $u_+^r(\boldsymbol{p})$ satisfies the equation

$$(\boldsymbol{\alpha}\boldsymbol{p} + m\beta)\, u_+^r(\boldsymbol{p}) = p_0 u_+^r(\boldsymbol{p}) \tag{6.17}$$

and is the eigenfunction of the spin projection operator in the direction of the momentum which has the eigenvalue r ($r = \pm 1$). We have

$$(u_+^{r'}(\boldsymbol{p}))^+ u_+^r(\boldsymbol{p}) = \delta_{r'r}. \tag{6.18}$$

The two other solutions of the Dirac equation

$$\frac{1}{(2\pi)^{3/2}} u_-^r(\boldsymbol{p})\, e^{i\boldsymbol{px}-ip_0x_0} \tag{6.19}$$

have negative energy ($-p_0$). The functions $u_-^r(\boldsymbol{p})$ satisfy the equation

$$(\boldsymbol{\alpha p}+m\beta)\, u_-^r(\boldsymbol{p}) = -p_0 u_-^r(\boldsymbol{p}). \tag{6.20}$$

It is obvious that

$$(u_+^{r'}(\boldsymbol{p}))^+ u_-^r(\boldsymbol{p}) = 0. \tag{6.21}$$

In addition

$$(u_-^{r'}(\boldsymbol{p}))^+ u_-^r(\boldsymbol{p}) = \delta_{r'r}. \tag{6.22}$$

The general solution of the Dirac equation can always be put in the form of a superposition of solutions (6.16) and (6.19). We have

$$\psi(x) = \frac{1}{(2\pi)^{3/2}} \int u_+^r(\boldsymbol{p})\, e^{i\boldsymbol{px}-ip_0x_0} \psi_+^r(\boldsymbol{p})\, d\boldsymbol{p} + \frac{1}{(2\pi)^{3/2}} \int u_-^r(\boldsymbol{p})\, e^{i\boldsymbol{px}-ip_0x_0} \psi_-^r(\boldsymbol{p})\, d\boldsymbol{p}. \tag{6.23}$$

Making the replacement $\boldsymbol{p} \to -\boldsymbol{p}$ in the second integral and introducing the notation

$$\psi_+^r(\boldsymbol{p}) = c_r(p), \quad \psi_-^r(-\boldsymbol{p}) = c_r(-p),$$

we obtain

$$\psi(x) = \frac{1}{(2\pi)^{3/2}} \int u_+^r(\boldsymbol{p})\, e^{ipx} c_r(p)\, d\boldsymbol{p} + \frac{1}{(2\pi)^{3/2}} \int u_-^r(-\boldsymbol{p})\, e^{-ipx} c_r(-p)\, d\boldsymbol{p},$$

where

$$px = p_\mu x_\mu = \boldsymbol{px} - p_0 x_0; \quad p_0 = +\sqrt{\boldsymbol{p}^2+m^2}. \tag{6.24}$$

Thus, the complex spinor field is characterized by four complex functions of the variable $\boldsymbol{p}$ [$c_r(p)$ and $c_r(-p)$].

Consider the case with $m \neq 0$. We introduce in place of $u_+^r(\boldsymbol{p})$ and $u_-^r(-\boldsymbol{p})$ the spinors

$$u^r(p) = \left(\frac{p_0}{m}\right)^{1/2} u_+^r(\boldsymbol{p}), \quad u^r(-p) = \left(\frac{p_0}{m}\right)^{1/2} u_-^r(-\boldsymbol{p}). \tag{6.25}$$

Using the relationship (see Appendix)

$$\bar{u}_\pm^{r'}(\boldsymbol{p})\, u_\pm^r(\boldsymbol{p}) = \pm\frac{m}{p_0} (u_\pm^{r'}(\boldsymbol{p}))^+ u_\pm^r(\boldsymbol{p}), \tag{6.26}$$

we obtain

$$\bar{u}^{r'}(p)\, u^r(p) = \delta_{r'r}, \quad \bar{u}^{r'}(-p)\, u^r(-p) = -\delta_{r'r}. \tag{6.27}$$

We further have

$$\bar{u}^{r'}(-p)\,u^r(p) = 0. \tag{6.27a}$$

In the case with $m \neq 0$ we shall use the spinors $u^r(p)$ and $u^r(-p)$ normalized by conditions (6.27) (invariant normalization). From (6.24) and (6.25) we obtain the following expansion for $\psi(x)$:

$$\psi(x) = \psi^{(+)}(x) + \psi^{(-)}(x), \tag{6.28}$$

where

$$\left.\begin{aligned} \psi^{(+)}(x) &= \frac{1}{(2\pi)^{3/2}} \int \left(\frac{m}{p_0}\right)^{1/2} u^r(p)\, e^{ipx}\, c_r(p)\, d\boldsymbol{p}, \\ \psi^{(-)}(x) &= \frac{1}{(2\pi)^{3/2}} \int \left(\frac{m}{p_0}\right)^{1/2} u^r(-p)\, e^{-ipx}\, c_r(-p)\, d\boldsymbol{p}. \end{aligned}\right\} \tag{6.28a}$$

Inserting this expansion into the expression (6.15) for the energy-momentum four-vector and integrating over $\boldsymbol{x}$ we obtain

$$\begin{aligned} P_\mu = \int & [c_r^*(p)\, c_{r'}(p')\, \bar{u}^r(p)\, \gamma_4 u^{r'}(p')\, \delta(\boldsymbol{p}'-\boldsymbol{p}) \\ & - c_r^*(p)\, c_{r'}(-p')\, \bar{u}^r(p)\, \gamma_4 u^{r'}(-p')\, e^{2ip_0x_0}\, \delta(\boldsymbol{p}'+\boldsymbol{p}) \\ & + c_r^*(-p)\, c_{r'}(p')\, \bar{u}^r(-p)\, \gamma_4 u^{r'}(p')\, e^{-2ip_0x_0}\, \delta(\boldsymbol{p}'+\boldsymbol{p}) \\ & - c_r^*(-p)\, c_{r'}(-p')\, \bar{u}^r(-p)\, \gamma_4 u^{r'}(-p')\, \delta(\boldsymbol{p}'-\boldsymbol{p})]\, p'_\mu \left(\frac{m^2}{p_0 p'_0}\right)^{1/2} d\boldsymbol{p}\, d\boldsymbol{p}'. \end{aligned}$$

Integrating over $\boldsymbol{p}'$ and recognizing that

$$\left.\begin{aligned} \bar{u}^r(p)\, \gamma_4 u^{r'}(p) &= \left(\frac{p_0}{m}\right) (u_+^r(\boldsymbol{p}))^+ u_+^{r'}(\boldsymbol{p}) = \frac{p_0}{m}\, \delta_{rr'}, \\ \bar{u}^r(-p)\, \gamma_4 u^{r'}(-p) &= \left(\frac{p_0}{m}\right) (u_-^r(-\boldsymbol{p}))^+ u_-^{r'}(-\boldsymbol{p}) = \frac{p_0}{m}\, \delta_{rr'}, \\ \bar{u}^r(p)\, \gamma_4 u^{r'}(-p')\big|_{\boldsymbol{p}'=-\boldsymbol{p}} &= \left(\frac{p_0}{m}\right) (u_+^r(\boldsymbol{p}))^+ u_-^{r'}(\boldsymbol{p}) = 0, \end{aligned}\right\} \tag{6.29}$$

we find for the energy and momentum of the field

$$H = \int [c_r^*(p)\, c_r(p) - c_r^*(-p)\, c_r(-p)]\, p_0\, d\boldsymbol{p}, \tag{6.30}$$

$$P_k = \int [c_r^*(p)\, c_r(p) - c_r^*(-p)\, c_r(-p)]\, p_k\, d\boldsymbol{p}. \tag{6.31}$$

Similarly we find that the charge of the complex spinor field is

$$Q = e \int [c_r^*(p)\, c_r(p) + c_r^*(-p)\, c_r(-p)]\, d\boldsymbol{p}. \tag{6.32}$$

The second term under the integral sign in (6.30) is negative (the contribution of states with negative energy). Thus, the energy of the classical spinor field is not a positive definite quantity. On the other hand, from (6.32) it follows that the charge of the classical complex spinor field can have only one sign.

We turn now to a discussion of the quantum spinor field. In expressions (6.28), (6.30), (6.31), and (6.32) we replace the functions $c_r(p)$ and $c_r(-p)$ with operators and the operation of complex conjugation of the classical quantities with the operation of Hermitian conjugation of operators. It is obvious that the operators of energy, momentum, and charge will be Hermitian operators after such a replacement.

The field operator $\psi(x)$ satisfies the free Dirac equation and is an operator in the interaction representation. Consequently, we have the relationship

$$i\frac{\partial\psi(x)}{\partial x_0} = [\psi(x), H]. \tag{6.33}$$

Insert the expansion (6.28) into (6.33):

$$\frac{1}{(2\pi)^{3/2}}\int\left\{e^{ipx}\left(\frac{m}{p_0}\right)^{1/2} u^r(p)(p_0 c_r(p) - [c_r(p), H]) + e^{-ipx}\left(\frac{m}{p_0}\right)^{1/2} u^r(-p)(-p_0 c_r(-p) - [c_r(-p), H])\right\} d\boldsymbol{p} = 0. \tag{6.34}$$

Hence,

$$\left.\begin{aligned}[c_r(p), H] &= p_0 c_r(p),\\ [c_r(-p), H] &= -p_0 c_r(-p).\end{aligned}\right\} \tag{6.35}$$

By Hermitian conjugation of (6.35) we also obtain

$$\left.\begin{aligned}[c_r^+(p), H] &= -p_0 c_r^+(p);\\ [c_r^+(-p), H] &= p_0 c_r^+(-p).\end{aligned}\right\} \tag{6.35a}$$

Denote by Φ_E the eigenvector of the Hamiltonian which has the eigenvalue E. We have

$$H\Phi_E = E\Phi_E. \tag{6.36}$$

Multiplying (6.36) by $c_r(-p)$ we obtain

$$c_r(-p)\,H\Phi_E = ([c_r(-p), H] + Hc_r(-p))\,\Phi_E = Ec_r(-p)\,\Phi_E.$$

From this, using (6.35) we find

$$Hc_r(-p)\,\Phi_E = (E+p_0)c_r(-p)\,\Phi_E.$$

Thus, the operator $c_r(-p)$ is the creation operator of a quantum with energy p_0. From (6.35) it is also clear that $c_r^+(p)$ is a creation operator and $c_r(p)$ and $c_r^+(-p)$ are annihilation operators. Denote

$$c_r(-p) = d_r^+(p).$$

For the operator $\psi(x)$ we obtain

$$\psi(x) = \psi^{(+)}(x) + \psi^{(-)}(x), \tag{6.37}$$

where

$$\left.\begin{aligned}\psi^{(+)}(x) &= \frac{1}{(2\pi)^{3/2}}\int\left(\frac{m}{p_0}\right)^{1/2} u^r(p)\,e^{ipx}c_r(p)\,d\boldsymbol{p};\\ \psi^{(-)}(x) &= \frac{1}{(2\pi)^{3/2}}\int\left(\frac{m}{p_0}\right)^{1/2} u^r(-p)\,e^{-ipx}\,d_r^+(p)\,d\boldsymbol{p}.\end{aligned}\right\}\tag{6.37a}$$

Hence,

$$\bar{\psi}(x) = \bar{\psi}^{(+)}(x)+\bar{\psi}^{(-)}(x),\tag{6.38}$$

where

$$\left.\begin{aligned}\bar{\psi}^{(+)}(x) &= \frac{1}{(2\pi)^{3/2}}\int\left(\frac{m}{p_0}\right)^{1/2} \bar{u}^r(-p)\,e^{ipx}\,d_r(p)\,d\boldsymbol{p},\\ \bar{\psi}^{(-)}(x) &= \frac{1}{(2\pi)^{3/2}}\int\left(\frac{m}{p_0}\right)^{1/2} \bar{u}^r(p)\,e^{-ipx}c_r^+(p)\,d\boldsymbol{p}.\end{aligned}\right\}\tag{6.38a}$$

Let Φ be an eigenvector of the Hamiltonian which describes some state of the quantized spinor field. From relationships (6.35) it follows that the vector $c_r^+(p)\Phi$ describes a state in which there exists an additional particle with four-momentum p. Furthermore, if the vector

$$c_r^+(p)c_r^+(p)\Phi\tag{6.39}$$

is not equal to zero, it is an eigenvector of the Hamiltonian and describes a state in which there exist two particles in the same quantum state. For particles with spin $\frac{1}{2}$, however, this is not possible. Pauli's principle requires that not more than one particle with spin $\frac{1}{2}$ be found in a given quantum state.

The commutation relations for the operators of the spinor field must be postulated therefore so that states such as (6.39) are forbidden. We shall assume that the operators of the non-Hermitian spinor field obey the following commutation relations:

$$\left.\begin{aligned}&[c_r(p),\,c_{r'}(p')]_+ = 0,\\ &[c_r(p),\,c_{r'}^+(p')]_+ = \delta_{rr'}\,\delta(\boldsymbol{p}-\boldsymbol{p}'),\\ &[d_r(p),\,d_{r'}(p')]_+ = 0,\\ &[d_r(p),\,d_{r'}^+(p')]_+ = \delta_{rr'}\,\delta(\boldsymbol{p}-\boldsymbol{p}'),\\ &[c_r(p),\,d_{r'}(p')]_+ = 0,\\ &[c_r(p),\,d_{r'}^+(p')]_+ = 0.\end{aligned}\right\}\tag{6.40}$$

Here $[A, B]_+ = AB+BA$ is an anticommutator. From (6.40) it follows that

$$\left.\begin{aligned}&[c_r^+(p),\,c_{r'}^+(p')]_+ = 0, \quad [d_r^+(p),\,d_{r'}^+(p')]_+ = 0,\\ &[c_r^+(p),\,d_{r'}^+(p')]_+ = 0, \quad [c_r^+(p),\,d_{r'}(p')]_+ = 0.\end{aligned}\right\}\tag{6.40a}$$

We shall show that the eigenfunctions of the Hamiltonian satisfy Pauli's principle. First we turn to the problem of the nature of the Hamiltonian of the quantized spinor field. If in the classical expression for the energy (6.30) we replace the quantities $c_r(p)$, $c_r^*(p)$, ... with the corresponding operators and by analogy with the preceding section write the field operators in normal order (placing the minus sign before the second term), then it is obvious that the resulting Hamiltonian can possess negative (arbitrarily large according to the model) eigenvalues. In addition, if we calculate the commutator $[d(p), H]$ using the commutation relations (6.40), we do not obtain relations (6.35). Define the Hamiltonian of the quantized spinor field in the following manner:

$$H = \int (c_r^+(p)\, c_r(p) + d_r^+(p)\, d_r(p))\, p_0\, d\boldsymbol{p}. \tag{6.41}$$

It is obvious that the eigenvalues of this Hamiltonian are positive. Let us calculate the commutator $[d_r(p), H]$. With the help of (6.40) we obtain

$$\left.\begin{aligned} [d_r(p), c_{r'}^+(p')\, c_{r'}(p')] &= [d_r(p), c_{r'}^+(p')]_+\, c_{r'}(p') = 0, \\ [d_r(p), d_{r'}^+(p')\, d_{r'}(p')] &= [d_r(p), d_{r'}^+(p')]_+\, d_{r'}(p') = \delta(\boldsymbol{p}-\boldsymbol{p}')\, d_r(p'). \end{aligned}\right\} \tag{6.42}$$

From (6.41) and (6.42) we find that $[d_r(p), H] = p_0 d_r(p)$. Similarly it is not difficult to verify that the Hamiltonian (6.41) also satisfies the remaining relations (6.35).

In order to formulate a procedure which would allow us to obtain the Hamiltonian (6.41) from the classical expression for the energy of the field (6.30), we assume that

$$\left.\begin{aligned} N(d_r(p)\, d_{r'}^+(p')) &= -d_{r'}^+(p')\, d_r(p), \\ N(c_{r'}^+(p')\, c_r(p)) &= c_{r'}^+(p')\, c_r(p). \end{aligned}\right\} \tag{6.43}$$

The operator N is defined in the following manner. When it acts on the product of two operators of the scalar (pseudoscalar) field, the operator N arranges them in normal order (the annihilation operator to the right of the creation operator); when it acts on the product of two operators of the spinor field, the operator N arranges them in normal order, and if there is a rearrangement of the field operators, the product of the operators written in normal order is multiplied by (-1). The definition of the operator N in the general case of n factors will be given later.

Let us assume that to obtain the operators of energy and all other physical quantities it is necessary to replace the functions in the corresponding classical expressions with field operators and to act on a product of field operators with the operator N. From the classical expression (6.30) we obtain the operator (6.41) and from (6.31) and (6.32) we find the following expressions for the momentum and charge operators:

$$P_k = \int (c_r^+(p)\, c_r(p) + d_r^+(p)\, d_r(p))\, p_k\, d\boldsymbol{p}, \tag{6.44}$$

$$Q = e \int (c_r^+(p)\, c_r(p) - d_r^+(p)\, d_r(p))\, d\boldsymbol{p}. \tag{6.45}$$

Let Φ_0 be the state vector with the lowest energy. Then

$$\left.\begin{aligned} c_r(p)\Phi_0 &= 0, \\ d_r(p)\Phi_0 &= 0. \end{aligned}\right\} \tag{6.46}$$

From (6.41), (6.44), (6.45), and (6.46) we find that

$$H\Phi_0 = 0, \quad \boldsymbol{P}\Phi_0 = 0, \quad Q\Phi_0 = 0.$$

Thus, Φ_0 is the vacuum state vector. It is clear that the vectors $c_r^+(p)\Phi_0$ and $d_r^+(p)\Phi_0$ describe a state with energy p_0. We shall show that these state vectors are also the eigenvectors of the momentum operator (6.44) and the charge operator (6.45). With the help of the commutation relations (6.40) we obtain

$$\left.\begin{aligned}[P, c_r^+(p)] &= \boldsymbol{p}c_r^+(p), & [Q, c_r^+(p)] &= ec_r^+(p),\\ [P, d_r^+(p)] &= \boldsymbol{p}d_r^+(p), & [Q, d_r^+(p)] &= -ed_r^+(p).\end{aligned}\right\} \tag{6.47}$$

Furthermore, using the standard procedure (replacing the product of the operators with the commutator plus the product of operators written in the reverse order), we easily find

$$\left.\begin{aligned}\boldsymbol{P}c_r^+(p)\Phi_0 &= \boldsymbol{p}c_r^+(p)\Phi_0,\\ \boldsymbol{P}d_r^+(p)\Phi_0 &= \boldsymbol{p}d_r^+(p)\Phi_0,\\ Qc_r^+(p)\Phi_0 &= ec_r^+(p)\Phi_0,\\ Qd_r^+(p)\Phi_0 &= -ed_r^+(p)\Phi_0.\end{aligned}\right\} \tag{6.48}$$

Thus, the state vector $c_r^+(p)\Phi_0$ describes the particle with four-momentum p, mass $m = \sqrt{p_0^2 - \boldsymbol{p}^2}$ and charge e and state vector $d_r^+(p)\Phi_0$ describes the antiparticle (the same mass, opposite charge) with four-momentum p.

We shall show that these vectors are also eigenvectors of the spin projection operator in the direction $\boldsymbol{p}$. We first obtain the expression for the spin of the field in classical theory. We choose $\psi(x)$ and $\bar{\psi}(x)$ as independent dynamic variables. Under the Lorentz transformations

$$x'_\mu = a_{\mu\varrho}x_\varrho$$

these spinors are transformed in the following manner:

$$\left.\begin{aligned}\psi'(x') &= L\psi(x),\\ \bar{\psi}'(x') &= \bar{\psi}(x)L^{-1},\end{aligned}\right\} \tag{6.49}$$

where the matrix L satisfies the condition

$$L^{-1}\gamma_\mu L = a_{\mu\varrho}\gamma_\varrho. \tag{6.50}$$

When the coefficients $a_{\mu\varrho}$ equal

$$a_{\mu\varrho} = \delta_{\mu\varrho} + \varepsilon_{\mu\varrho}, \tag{6.51}$$

where $\varepsilon_{\mu\varrho} = -\varepsilon_{\varrho\mu}$ are infinitesimal quantities [see (3.18) and following], accurate to terms linear in ε, we have

$$\begin{aligned}L &= 1 + \tfrac{1}{2}\boldsymbol{\Sigma}^{\mu\varrho}\varepsilon_{\mu\varrho},\\ L^{-1} &= 1 - \tfrac{1}{2}\boldsymbol{\Sigma}^{\mu\varrho}\varepsilon_{\mu\varrho}.\end{aligned} \tag{6.52}$$

Inserting (6.51) and (6.52) into (6.50) we find that the matrix $\boldsymbol{\Sigma}^{\mu\varrho}$ must obey the relationship

$$[\gamma_\nu, \boldsymbol{\Sigma}^{\mu\varrho}] = \delta_{\nu\mu}\gamma_\varrho - \delta_{\nu\varrho}\gamma_\mu. \tag{6.53}$$

Using the commutation relations for the matrices γ_μ, we easily verify that (6.53) is satisfied if

$$\boldsymbol{\Sigma}^{\mu\varrho} = \tfrac{1}{4}(\gamma_\mu\gamma_\varrho - \gamma_\varrho\gamma_\mu) = i\,\tfrac{1}{2}\sigma_{\mu\varrho}. \tag{6.54}$$

Here

$$\sigma_{\mu\varrho} = \frac{1}{2i}(\gamma_\mu\gamma_\varrho - \gamma_\varrho\gamma_\mu). \tag{6.55}$$

From (6.49) and (6.52) we find that

$$\left.\begin{aligned} \psi'(x') &= (1+\tfrac{1}{2}\boldsymbol{\Sigma}^{\mu\varrho}\varepsilon_{\mu\varrho})\,\psi(x), \\ \bar{\psi}'(x') &= (1-\tfrac{1}{2}\tilde{\boldsymbol{\Sigma}}^{\mu\varrho}\varepsilon_{\mu\varrho})\,\bar{\psi}(x). \end{aligned}\right\} \tag{6.56}$$

Repeating the discussion of the third section and taking (6.56) into account we obtain for the complex spinor field the following expression for the tensor $S_{\mu\varrho}$:

$$S_{\mu\varrho} = i\int \frac{\partial\mathcal{L}}{\partial\dfrac{\partial\psi_{\sigma'}}{\partial x_4}}\boldsymbol{\Sigma}^{\mu\varrho}_{\sigma'\sigma}\psi_\sigma\,d\boldsymbol{x} - i\int\frac{\partial\mathcal{L}}{\partial\dfrac{\partial\bar{\psi}_{\sigma'}}{\partial x_4}}\tilde{\boldsymbol{\Sigma}}^{\mu\varrho}_{\sigma'\sigma}\bar{\psi}_\sigma\,d\boldsymbol{x}. \tag{6.57}$$

From (6.7), (6.8), and (6.57) we find that

$$S_{\mu\varrho} = \tfrac{1}{4}\int\bar{\psi}[\gamma_4\sigma_{\mu\varrho}+\sigma_{\mu\varrho}\gamma_4]\,\psi\,d\boldsymbol{x}. \tag{6.58}$$

Hence

$$S_{ik} = \tfrac{1}{2}\int\psi^+\sigma_{ik}\psi\,d\boldsymbol{x}, \qquad S_{i4} = S_{4i} = 0 \tag{6.59}$$

(i, k take values 1, 2, 3).

Define the spin vector of the field by the expression

$$S_i = \tfrac{1}{2}e_{ikl}S_{kl}. \tag{6.60}$$

Using (6.59) and (6.55) we obtain

$$S_i = \tfrac{1}{2}\int\psi^+\boldsymbol{\Sigma}_i\psi\,d\boldsymbol{x}, \tag{6.61}$$

where the matrices $\boldsymbol{\Sigma}_i$ are equal to

$$\boldsymbol{\Sigma}_i = \tfrac{1}{2}e_{ikl}\sigma_{kl} = \frac{1}{2i}e_{ikl}\gamma_k\gamma_l = \frac{1}{2i}e_{ikl}\alpha_k\alpha_l. \tag{6.62}$$

It is obvious that the $\boldsymbol{\Sigma}_i$ are Hermitian matrices. From the commutation relations for the matrices γ_i it is not difficult to see that the matrices $\frac{1}{2}\boldsymbol{\Sigma}_i$ satisfy the commutation relations of the angular momentum operator

$$[\boldsymbol{\Sigma}_i, \boldsymbol{\Sigma}_k] = 2ie_{ikl}\boldsymbol{\Sigma}_l. \tag{6.63}$$

Note that in the Dirac–Pauli representation the 4×4 matrices Σ_i are equal to

$$\boldsymbol{\Sigma}_i = \begin{pmatrix}\sigma_i & 0\\ 0 & \sigma_i\end{pmatrix}, \tag{6.64}$$

where σ_i are the 2×2 Pauli matrices.

Replacing $\psi(x)$ and $\psi^+(x)$ in expressions (6.61) with operators and acting on the product of the field operators with the operator N, we obtain the spin operator of the quantized spinor field. Let us calculate the commutators

$$[c_r(p),\ \boldsymbol{Sn}],\quad [d_r(p),\ \boldsymbol{Sn}]. \tag{6.65}$$

Here $\boldsymbol{n} = \boldsymbol{p}/|\boldsymbol{p}|$ is the unit vector in the direction of the momentum $\boldsymbol{p}$. Inserting the expansions (6.37) and (6.38) into the expression for the spin operator, then calculating the commutators of the field operators, and carrying out the corresponding integrations, we obtain

$$[c_r(p),\ \boldsymbol{Sn}] = \tfrac{1}{2}\left((u_+^r(\boldsymbol{p}))^+ \boldsymbol{\Sigma n} u_+^{r'}(\boldsymbol{p})\right) c_{r'}(p) + \tfrac{1}{2}\, e^{2ip_0x_0}\left((u_+^r(\boldsymbol{p}))^+ \boldsymbol{\Sigma n} u_-^{r'}(\boldsymbol{p})\right) d_{r'}^+(p')|_{\boldsymbol{p}'=-\boldsymbol{p}}. \tag{6.66}$$

The spinors $u_+^r(\boldsymbol{p})$ and $u_-^r(\boldsymbol{p})$ are eigenfunctions of the operator $(\boldsymbol{\Sigma n})$:

$$\left.\begin{aligned} \boldsymbol{\Sigma n} u_+^r(\boldsymbol{p}) &= r u_+^r(\boldsymbol{p}), \\ \boldsymbol{\Sigma n} u_-^r(\boldsymbol{p}) &= r u_-^r(\boldsymbol{p}). \end{aligned}\right\} \tag{6.67}$$

From (6.66) and (6.67), taking into account the conditions of normalization (6.18) and orthogonality (6.21), we obtain

$$[c_r(p),\ \boldsymbol{Sn}] = \tfrac{1}{2} r c_r(p). \tag{6.68}$$

Similarly we find that

$$\begin{aligned} [d_r(p),\ \boldsymbol{Sn}] = &-\tfrac{1}{2}\, e^{2ip_0x_0}\left((u_+^{r'}(-\boldsymbol{p}))^+ \boldsymbol{\Sigma n} u_-^r(-\boldsymbol{p})\right) c_{r'}^+(p')|_{\boldsymbol{p}'=-\boldsymbol{p}} \\ &-\tfrac{1}{2}\left((u_-^{r'}(-\boldsymbol{p}))^+ \boldsymbol{\Sigma n} u_-^r(-\boldsymbol{p})\right) d_{r'}(p). \end{aligned} \tag{6.69}$$

From this with the help of (6.67), (6.21), and (6.22) we obtain

$$[d_r(p),\ \boldsymbol{Sn}] = \tfrac{1}{2} r d_r(p). \tag{6.70}$$

From (6.68) and (6.70) we find

$$\left.\begin{aligned} [\boldsymbol{Sn},\ c_r^+(p)] &= \tfrac{1}{2} r c_r^+(p), \\ [\boldsymbol{Sn},\ d_r^+(p)] &= \tfrac{1}{2} r d_r^+(p). \end{aligned}\right\} \tag{6.71}$$

Acting with the operator $\boldsymbol{Sn}$ on the state vectors $c_r^+(p)\Phi_0$ and $d_r^+(p)\Phi_0$, using the customary method, and recognizing that $\boldsymbol{Sn}\Phi_0 = 0$, we obtain

$$\left.\begin{aligned} \boldsymbol{Sn} c_r^+(p)\,\Phi_0 = ([\boldsymbol{Sn},\ c_r^+(p)] + c_r^+(p)\,\boldsymbol{Sn})\Phi_0 &= \tfrac{1}{2} r c_r^+(p)\,\Phi_0, \\ \boldsymbol{Sn}\, d_r^+(p)\Phi_0 &= \tfrac{1}{2} r d_r^+(p)\,\Phi_0. \end{aligned}\right\} \tag{6.72}$$

Thus, the state vector $c_r^+(p)\Phi_0$ describes a particle with charge e, four-momentum p, and spin projection in the direction of momentum equal to $r/2$ and the vector $d_r^+(p)\Phi_0$ describes an antiparticle with charge $-e$, four-momentum p, and spin projection in the direction of momentum $r/2$.

We continue now the examination of the eigenvectors of the Hamiltonian. It is obvious that the state vector $c_{r_2}^+(p_2)c_{r_1}^+(p_1)\Phi_0$ is an eigenvector of the Hamiltonian (6.41) as well as an eigenvector of the momentum and charge operators. This vector describes two particles with four-momenta p_1 and p_2. The state vector $d_{r'}^+(p')\, c_r^+(p)\Phi_0$ describes a particle with four-momentum p and an antiparticle with four-momentum p'. Thus, if we act on the vacuum state vector with the creation operators c^+ and d^+, we obtain the eigenvectors of the Hamiltonian (6.41) which describe free particles and antiparticles in states with definite momenta.

Consider, for example, the state vector

$$d_{r'}^{+}(p')\, c_{r_1}^{+}(p_1)\, c_{r_2}^{+}(p_2)\, c_{r_3}^{+}(p_3)\, \Phi_0, \tag{6.73}$$

which describes three particles and one antiparticle. Using the commutation relations (6.40) we obtain

$$d_{r'}^{+}(p')\, c_{r_1}^{+}(p_1)\, c_{r_2}^{+}(p_2)\, c_{r_3}^{+}(p_3)\, \Phi_0 = -d_{r'}^{+}(p')\, c_{r_3}^{+}(p_3)\, c_{r_2}^{+}(p_2)\, c_{r_1}^{+}(p_1)\, \Phi_0.$$

If we assume that $p_1 = p_3$ and $r_1 = r_3$, the vector (6.73) vanishes. It is clear that in the general case the state vector vanishes if at least two particle creation operators (or two antiparticle creation operators) possess identical p and r. Consequently, the commutation relations for the operators of the spinor field are postulated in accordance with the Pauli principle (not more than one particle can be found in a state with a given momentum and spin projection).

7. The Electromagnetic Field

A free electromagnetic field is described by the potential $A_\mu(x)$ which satisfies the equation

$$\Box A_\mu = 0 \tag{7.1}$$

and the Lorentz condition

$$\frac{\partial A_\mu}{\partial x_\mu} = 0. \tag{7.2}$$

Equation (7.1) with the condition that the potential satisfies (7.2) is equivalent to the free Maxwell equations

$$\frac{\partial F_{\mu\nu}}{\partial x_\nu} = 0, \tag{7.3}$$

where $F_{\mu\nu} = \partial A_\nu/\partial x_\mu - \partial A_\mu/\partial x_\nu$ is the field strength tensor. If we replace the potential A_μ with

$$A'_\mu = A_\mu + \frac{\partial \Lambda}{\partial x_\mu}, \tag{7.4}$$

where Λ is an arbitrary function satisfying the equation

$$\Box \Lambda = 0, \tag{7.5}$$

then

$$\Box A'_\mu = 0, \quad \frac{\partial A'_\mu}{\partial x_\mu} = 0. \tag{7.6}$$

Thus, the potential of an electromagnetic field is not defined uniquely by equations (7.1) and (7.2). The potentials A_μ and A'_μ give, however, the same field strength tensor (gradient invariance).

In § 2 it was shown that the Lagrangian of a free electromagnetic field can be chosen to be

$$\mathcal{L} = -\tfrac{1}{4}F_{\mu\nu}F_{\mu\nu}. \tag{7.7}$$

It is not difficult to see that

$$-\tfrac{1}{4}F_{\mu\nu}F_{\mu\nu} = -\tfrac{1}{2}\frac{\partial A_\nu}{\partial x_\mu}\frac{\partial A_\nu}{\partial x_\mu} + \tfrac{1}{2}\frac{\partial}{\partial x_\mu}\left(A_\nu\frac{\partial A_\mu}{\partial x_\nu}\right) - \tfrac{1}{2}A_\nu\frac{\partial}{\partial x_\nu}\left(\frac{\partial A_\mu}{\partial x_\mu}\right). \tag{7.8}$$

The last term in this expression is zero due to the Lorentz condition. The term

$$\frac{\partial}{\partial x_\mu}\left(A_\nu\frac{\partial A_\mu}{\partial x_\nu}\right)$$

does not contribute to the variation of the action. Consequently, the Lagrangians (7.7) and

$$\mathcal{L} = -\frac{1}{2}\frac{\partial A_\mu}{\partial x_\nu}\frac{\partial A_\mu}{\partial x_\nu} \tag{7.9}$$

imply the same field equations. We shall take for the Lagrangian of the free electromagnetic field expression (7.9). Using the general formulae of § 3 we find from (7.9) that the energy-momentum tensor of the electromagnetic field is

$$T_{\mu\nu} = \frac{\partial\mathcal{L}}{\partial\dfrac{\partial A_\varrho}{\partial x_\nu}}\frac{\partial A_\varrho}{\partial x_\mu} - \mathcal{L}\delta_{\mu\nu} = -\frac{\partial A_\varrho}{\partial x_\nu}\cdot\frac{\partial A_\varrho}{\partial x_\mu} + \frac{1}{2}\frac{\partial A_\varrho}{\partial x_\sigma}\frac{\partial A_\varrho}{\partial x_\sigma}\delta_{\mu\nu}. \tag{7.10}$$

For the energy-momentum vector we obtain

$$P_\mu = i\int T_{\mu 4}\,dx = i\int\left[-\frac{\partial A_\varrho}{\partial x_4}\frac{\partial A_\varrho}{\partial x_\mu} + \frac{1}{2}\frac{\partial A_\varrho}{\partial x_\sigma}\frac{\partial A_\varrho}{\partial x_\sigma}\delta_{4\mu}\right]dx. \tag{7.11}$$

Let us expand the potential $A_\mu(x)$ in plane waves. Keeping in mind the equation of the field (7.1) we obtain

$$A_\mu(x) = \frac{1}{(2\pi)^{3/2}}\int\frac{1}{\sqrt{2k_0}}\,e^{ikx-ik_0x_0}A_\mu^{(+)}(\boldsymbol{k})\,d\boldsymbol{k} + \frac{1}{(2\pi)^{3/2}}\int\frac{1}{\sqrt{2k_0}}\,e^{ikx+ik_0x_0}A_\mu^{(-)}(\boldsymbol{k})\,d\boldsymbol{k}, \tag{7.12}$$

where $k_0 = |\boldsymbol{k}|$. The first term is the positive-frequency part of the potential and the second term is the negative-frequency part. Replacing in the second integral $\boldsymbol{k} \to -\boldsymbol{k}$, we rewrite expansion (7.12) in the form

$$A_\mu(x) = \frac{1}{(2\pi)^{3/2}}\int\frac{1}{\sqrt{2k_0}}\left[A_\mu^{(+)}(\boldsymbol{k})\,e^{ikx} + A_\mu^{(-)}(-\boldsymbol{k})\,e^{-ikx}\right]d\boldsymbol{k}. \tag{7.12a}$$

Define the four-vectors

$$e^\lambda(k) = (\boldsymbol{e}_\lambda(\boldsymbol{k}), 0), \quad e^4(k) = (0, i), \quad \lambda = 1, 2, 3. \tag{7.13}$$

Here $\boldsymbol{e}_3(\boldsymbol{k}) = \boldsymbol{k}/k_0 = \boldsymbol{\varkappa}$ is the unit vector directed along $\boldsymbol{k}$ and $\boldsymbol{e}_1(\boldsymbol{k})$ and $\boldsymbol{e}_2(\boldsymbol{k})$ are unit vectors orthogonal to one another and to vector $\boldsymbol{k}$. We have

$$\boldsymbol{e}_\lambda(\boldsymbol{k})\,\boldsymbol{e}_{\lambda'}(\boldsymbol{k}) = \delta_{\lambda\lambda'}, \qquad \lambda,\ \lambda' = 1,\ 2,\ 3. \tag{7.14}$$

It is obvious that the four-vectors $e^\lambda(k)$ satisfy the conditions

$$(e^\lambda(k)\,e^{\lambda'}(k)) = \delta_{\lambda\lambda'}\eta_\lambda, \qquad (\lambda = 1,\ 2,\ 3,\ 4) \tag{7.15}$$

$$(e^\lambda(k)\,k) = 0, \quad (\lambda = 1,\ 2), \qquad (e^3(k)\,k) = k_0, \qquad (e^4(k)\,k) = -k_0. \tag{7.16}$$

The factor η_λ is defined in the following manner:

$$\eta_\lambda = \begin{cases} 1, & \lambda = 1,\ 2,\ 3, \\ -1, & \lambda = 4. \end{cases} \tag{7.17}$$

Note that in (7.15) there is no summation over the repeated indices λ.

The coefficients $A_\mu^{(+)}(\boldsymbol{k})$ and $A_\mu^{(-)}(-\boldsymbol{k})$ in expression (7.12) can be expanded in the complete set of vectors $e^\lambda(k)$

$$\left.\begin{aligned} A_\mu^{(+)}(\boldsymbol{k}) &= \sum_{\lambda=1}^{4} a_\lambda(k)\,e_\mu^\lambda(k), \\ A_\mu^{(-)}(-\boldsymbol{k}) &= \sum_{\lambda=1}^{4} a_\lambda(-k)\,e_\mu^\lambda(k). \end{aligned}\right\} \tag{7.18}$$

From (7.13) and (7.18) it follows that

$$\left.\begin{aligned} a_\lambda(k) &= (e^\lambda(k)\,A^{(+)}(\boldsymbol{k})) = (\boldsymbol{e}_\lambda(\boldsymbol{k})\,\boldsymbol{A}^{(+)}(\boldsymbol{k})), \qquad \lambda = 1,\ 2, \\ a_3(k) &= (\boldsymbol{\varkappa}\boldsymbol{A}^{(+)}(\boldsymbol{k})), \\ a_4(k) &= -iA_4^{(+)}(\boldsymbol{k}) = A_0^{(+)}(\boldsymbol{k}). \end{aligned}\right\} \tag{7.19}$$

Thus, $a_3(k)$ is the projection of the vector $\boldsymbol{A}^{(+)}(\boldsymbol{k})$ in the direction of the vector $\boldsymbol{k}$ (the longitudinal component of the vector $\boldsymbol{A}^{(+)}(k)$); $a_1(k)$ and $a_2(k)$ are the projections of the vector $\boldsymbol{A}^{(+)}(\boldsymbol{k})$ on the vectors perpendicular to $\boldsymbol{k}$ (the transverse components of $\boldsymbol{A}^{(+)}(\boldsymbol{k})$), and $a_4(k)$ is the Fourier transform of the scalar potential $A_0^{(+)}(x)$. Similarly we find that $a_1(-k)$, $a_2(-k)$, and $a_3(-k)$ are the transverse and longitudinal components of the vector $\boldsymbol{A}^{(-)}(-\boldsymbol{k})$ and $a_4(-k) = A_0^{(-)}(-\boldsymbol{k})$.

Inserting expansions (7.18) into (7.12a) we obtain

$$A_\mu(x) = A_\mu^{(+)}(x) + A_\mu^{(-)}(x) = \frac{1}{(2\pi)^{3/2}} \int \frac{1}{\sqrt{2k_0}} \sum_{\lambda=1}^{4} e_\mu^\lambda(k)\,[a_\lambda(k)\,e^{ikx} + a_\lambda(-k)\,e^{-ikx}]\,d\boldsymbol{k}. \tag{7.20}$$

The space part of the four-vector A_μ is real and the time part is imaginary. Thus

$$A_\mu^* = A_\mu \eta_\mu \tag{7.21}$$

(there is no summation over μ). Inserting expansion (7.20) into (7.21) and recognizing that

$$(e_\mu^\lambda(k))^* = e_\mu^\lambda(k)\,\eta_\mu,$$

we obtain

$$a_\lambda(-k) = a_\lambda^*(k). \tag{7.22}$$

Finally we find the following expansion for the potential:

$$A_\mu(x) = \frac{1}{(2\pi)^{3/2}} \int \frac{1}{\sqrt{2k_0}} \sum_{\lambda=1}^{4} e_\mu^\lambda(k)[a_\lambda(k)\,e^{ikx} + a_\lambda^*(k)\,e^{-ikx}]\,d\boldsymbol{k}. \tag{7.23}$$

Let us now consider what relationships between the $a_\lambda(k)$ result from the Lorentz condition. We have

$$\frac{\partial A_\mu}{\partial x_\mu} = 0 = \frac{i}{(2\pi)^{3/2}} \int \frac{1}{\sqrt{2k_0}} \sum_{\lambda=1}^{4} (e^\lambda(k)\,k)\,[a_\lambda(k)\,e^{ikx} - a_\lambda^*(k)\,e^{-ikx}]\,d\boldsymbol{k}. \tag{7.24}$$

From (7.24) it follows that

$$\sum_{\lambda=1}^{4} (e^\lambda(k)\,k)\,a_\lambda(k) = 0, \tag{7.25}$$

from which, taking account of (7.16), we obtain

$$a_3(k) = a_4(k). \tag{7.26}$$

Thus, the electromagnetic field can be described by the complex functions $a_\lambda(k)$—functions of the variable $\boldsymbol{k}$ (the Fourier components of the potential) which as a result of (7.2) satisfy relationship (7.26). Let us insert the expansion (7.23) into the expression for the energy of the field. Doing the corresponding integrations and using (7.15) we find

$$H = \int \sum_{\lambda,\lambda'=1}^{4} a_\lambda^*(k)\,a_{\lambda'}(k)\,(e^\lambda(k)\,e^{\lambda'}(k))\,k_0\,d\boldsymbol{k} = \int \sum_{\lambda=1}^{4} a_\lambda^*(k)\,a_\lambda(k)\eta_\lambda k_0\,d\boldsymbol{k}. \tag{7.27}$$

The terms $a_3^*(k)\,a_3(k)$ and $a_4^*(k)\,a_4(k)$ (the contributions of the longitudinal and scalar components) which appear in (7.27) with different signs cancel as a result of (7.26) and finally we obtain the following expression for the energy of the free electromagnetic field:

$$H = \int \sum_{\lambda=1,2} a_\lambda^*(k)\,a_\lambda(k)\,k_0\,d\boldsymbol{k}. \tag{7.28}$$

Obviously H can assume only positive values. Note that the energy of the field becomes a positive definite quantity only after the additional Lorentz condition is applied.

For the momentum of the free electromagnetic field we obtain the expression

$$P_i = \int \sum_{\lambda=1,2} a_\lambda^*(k)\,a_\lambda(k)\,k_i\,d\boldsymbol{k}. \tag{7.29}$$

Thus, only the transverse components of the potential contribute to the energy and momentum of the field. On this we conclude the examination of the classical electromagnetic field and now turn to quantum theory.

Consider $a_\lambda(k)$ and $a_\lambda(-k)$ as operators. Then $A_\mu(x)$ is also an operator. The operator $A_\mu(x)$ is given by

$$A_\mu(x) = \frac{1}{(2\pi)^{3/2}} \int \frac{1}{\sqrt{2k_0}} \sum_{\lambda=1}^{4} e_\mu^\lambda(k)\,[a_\lambda(k)\,e^{ikx} + a_\lambda(-k)\,e^{-ikx}]\,d\boldsymbol{k}. \tag{7.30}$$

From (7.30) it is obvious that the operator $A_\mu(x)$ satisfies the equation

$$\Box A_\mu(x) = 0. \tag{7.31}$$

In the classical case the potential satisfies the conditions (7.21). In accordance with this we require that the operator $\boldsymbol{A}(x)$ be Hermitian $[\boldsymbol{A}^+(x) = \boldsymbol{A}(x)]$ and the operator $A_4(x)$ be anti-Hermitian $[A_4^+(x) = -A_4(x)]$. These conditions of the operator $A_\mu(x)$ can be written

$$A_\mu^+(x) = A_\mu(x)\,\eta_\mu \tag{7.32}$$

(no summation over μ). Inserting expansion (7.30) into (7.32) we find that

$$a_\lambda(-k) = a_\lambda^+(k). \tag{7.33}$$

Thus,

$$A_\mu(x) = A_\mu^{(+)}(x) + A_\mu^{(-)}(x), \tag{7.34}$$

where

$$\left.\begin{aligned} A_\mu^{(+)}(x) &= \frac{1}{(2\pi)^{3/2}} \int \frac{1}{\sqrt{2k_0}} \sum_{\lambda=1}^{4} a_\lambda(k)\, e_\mu^\lambda(k)\, \mathrm{e}^{ikx}\, d\boldsymbol{k}, \\ A_\mu^{(-)}(x) &= \frac{1}{(2\pi)^{3/2}} \int \frac{1}{\sqrt{2k_0}} \sum_{\lambda=1}^{4} a_\lambda^+(k)\, e_\mu^\lambda(k)\, \mathrm{e}^{-ikx}\, d\boldsymbol{k}. \end{aligned}\right\} \tag{7.34a}$$

Furthermore, $A_\mu(x)$ is an operator in the interaction representation and satisfies therefore the equation

$$i\,\frac{\partial A_\mu(x)}{\partial x_0} = [A_\mu(x), H], \tag{7.35}$$

where H is the Hamiltonian of the free electromagnetic field. Inserting the expansion (7.34) into (7.35) we obtain

$$k_0 a_\lambda(k) = [a_\lambda(k), H], \tag{7.36}$$

$$-k_0 a_\lambda^+(k) = [a_\lambda^+(k), H]. \tag{7.37}$$

Using expressions (7.11) and (7.34) we find for the energy and momentum operators of the quantized electromagnetic field

$$H = \int \sum_{\lambda=1}^{4} a_\lambda^+(k)\, a_\lambda(k)\, \eta_\lambda k_0\, d\boldsymbol{k}, \tag{7.38}$$

$$P_i = \int \sum_{\lambda=1}^{4} a_\lambda^+(k)\, a_\lambda(k)\, \eta_\lambda k_i\, d\boldsymbol{k}. \tag{7.39}$$

We shall now discuss the commutation relations for the field operators. Suppose that the operators $a_\lambda(k)$ and $a_\lambda(k')$ commute, i.e.

$$[a_\lambda(k), a_{\lambda'}(k')] = 0. \tag{7.40a}$$

From (7.40a) we obtain

$$[a_\lambda^+(k), a_{\lambda'}^+(k')] = 0. \tag{7.40b}$$

Using (7.38) and (7.40) we find from (7.36)

$$k_0 a_\lambda(k) = \int \sum_{\lambda'=1}^{4} [a_\lambda(k),\ a^+_{\lambda'}(k')]\, a_{\lambda'}(k')\, \eta_{\lambda'} k'_0\, d\boldsymbol{k}'. \tag{7.41}$$

Hence,

$$[a_\lambda(k),\ a^+_{\lambda'}(k')] = \delta(\boldsymbol{k}-\boldsymbol{k}')\, \delta_{\lambda\lambda'}\eta_\lambda. \tag{7.40c}$$

Thus, we postulate the commutation relations (7.40) for the operators of the electromagnetic field. Note the significant difference between the commutation relations for $a_\lambda(k)$ and $a^+_{\lambda'}(k')$ with $\lambda,\ \lambda' = 1, 2, 3$ and the commutation relations for $a_4(k)$ and $a^+_4(k')$ [the minus sign on the right-hand side of (7.40c) is for $\lambda = \lambda' = 4$]. This difference is related to the fact that the term $a^+_4(k)\, a_4(k)$ enters the Hamiltonian with a minus sign.

Using commutation relations (7.40) we obtain

$$[a_\lambda(k),\ P_i] = k_i a_\lambda(k). \tag{7.41a}$$

From this we also find

$$[a^+_\lambda(k),\ P_i] = -k_i a^+_\lambda(k). \tag{7.41b}$$

We turn now to the formulation of the Lorentz condition in quantum theory. For the classical quantities the relationship (7.26) results from the Lorentz condition. In quantization all the operators $a_\lambda(k)$ are independent. In order to ensure that the Lorentz condition is satisfied by the matrix elements, we require

$$\frac{\partial A^{(+)}_\mu}{\partial x_\mu}\Phi = 0, \tag{7.42a}$$

where Φ is any vector describing the state of the quantized electromagnetic field.

Hence

$$\Phi^+_1 \left(\frac{\partial A^{(+)}_\mu}{\partial x_\mu}\right)^+ = \Phi^+_1 \frac{\partial A^{(-)}_\mu}{\partial x_\mu} = 0. \tag{7.42b}$$

Adding (7.42a) and (7.42b) we obtain

$$\left(\Phi^+_1 \frac{\partial A_\mu}{\partial x_\mu}\Phi\right) = 0. \tag{7.43}$$

Thus, if condition (7.42a) is satisfied for all state vectors of the electromagnetic field, the matrix elements of the operator $\partial A_\mu/\partial x_\mu$ are zero. From (7.43) and (7.31) it follows that

$$\frac{\partial}{\partial x_\nu}(\Phi^+_1 F_{\mu\nu}(x)\Phi) = 0. \tag{7.44}$$

Inserting expansion (7.35) into (7.42) we conclude that the vectors Φ which describe the state of a quantized electromagnetic field must satisfy the conditions

$$\sum_{\lambda=1}^{4} (e^\lambda k)\, a_\lambda(k)\Phi = k_0(a_3(k) - a_4(k))\Phi = 0 \tag{7.45a}$$

($\boldsymbol{k}$ is an arbitrary vector). Hence

$$\Phi^+(a_3^+(k) - a_4^+(k)) = 0. \tag{7.45b}$$

We shall now construct a set of vectors which describe the possible states of the quantized electromagnetic field. From the postulated commutation relations (7.40) it follows that there is no state vector Φ for which $a_4(k)\Phi = 0$. Actually, let us examine the commutation relation

$$[a_4(k), a_4^+(k')] = -\delta(\boldsymbol{k} - \boldsymbol{k}').$$

Multiply it on the left by $\Phi^+ h^*(\boldsymbol{k})$ and on the right by $h(\boldsymbol{k}')\Phi$ and integrate over $\boldsymbol{k}$ and $\boldsymbol{k}'$ [$h(\boldsymbol{k})$ is an arbitrary function of $\boldsymbol{k}$, Φ is some state vector]. Then

$$\begin{aligned}&\left(\int h(\boldsymbol{k})\, a_4^+(k)\, d\boldsymbol{k}\, \Phi\right)^+ \left(\int h(\boldsymbol{k})\, a_4^+(k)\, d\boldsymbol{k}\, \Phi\right)\\ &\quad - \left(\int h^*(\boldsymbol{k})\, a_4(k)\, d\boldsymbol{k}\, \Phi\right)^+ \left(\int h^*(\boldsymbol{k})\, a_4(k)\, d\boldsymbol{k}\, \Phi\right) = -(\Phi^+\Phi) \int |h(\boldsymbol{k})|^2\, d\boldsymbol{k}.\end{aligned} \tag{7.46}$$

If there did exist a state vector Φ with $a_4(k)\Phi = 0$, then for such a state the second term on the left-hand side of (7.46) would disappear and we would arrive at a contradiction [the left side of (7.46) is a positive quantity and the right is negative]. Thus, in our scheme of quantization for any vector Φ

$$a_4(k)\Phi \neq 0. \tag{7.47}$$

From the Lorentz condition (7.45) it also follows that for all allowable state vectors

$$a_3(k)\Phi \neq 0. \tag{7.48}$$

Now let us find the mean value of the energy and momentum operators in the state described by some vector Φ. Using expressions (7.38) and (7.39) and isolating terms with $\lambda = 3$ and $\lambda = 4$, we have

$$\begin{aligned}\langle P_\mu \rangle = (\Phi^+ P_\mu \Phi) &= \left(\Phi^+ \int \sum_{\lambda=1,2} a_\lambda^+(k)\, a_\lambda(k)\, k_\mu\, d\boldsymbol{k}\, \Phi\right)\\ &+ \left(\Phi^+ \int [a_3^+(k)\, a_3(k) - a_4^+(k)\, a_4(k)]\, k_\mu\, d\boldsymbol{k}\, \Phi\right).\end{aligned} \tag{7.49}$$

Using the Lorentz condition (7.45) we obtain

$$(\Phi^+ a_4^+(k)\, a_4(k)\Phi) = (\Phi^+ a_3^+(k)\, a_3(k)\Phi).$$

The second term on the right-hand side of (7.49) is therefore equal to zero; as a result we have for the mean value of the energy-momentum operator

$$\langle P_\mu \rangle = \left(\Phi^+ \int \sum_{\lambda=1,2} a_\lambda^+(k)\, a_\lambda(k)\, k_\mu\, d\boldsymbol{k}\, \Phi\right). \tag{7.50}$$

Thus, only photons with transverse polarization [$e^1(k)$ and $e^2(k)$] contribute to the energy and momentum of an electromagnetic field. Obviously that state in which photons with transverse polarization are absent has the least energy. Let Φ_0 be a state vector satisfying the conditions

$$a_\lambda(k)\Phi_0 = 0, \quad \lambda = 1, 2. \tag{7.51}$$

From (7.50) we conclude that the energy and momentum of the field in the state Φ_0 are zero (Φ_0 is the vacuum state). The condition (7.51) means that in the state described by Φ_0 there are no photons with transverse polarization.

Acting on the vector Φ_0 with the creation operators $a_\lambda^+(k)$, one can construct a set of state vectors describing free photons with definite four-momenta.

Our goal is to calculate the matrix elements for transitions between states of free particles —the matrix elements of the S-matrix, which [see (1.38)] are defined by the interaction Hamiltonian. The interaction Hamiltonian of an electromagnetic field with other fields can contain the operators $A_\mu(x)$ and $F_{\mu\nu}(x)$. The matrix elements of the product of the field operators $A(x)$ and their derivatives will be calculated, consequently, in finding the corresponding matrix elements of the S-matrix. It will be shown in the following chapter that the calculation of S-matrix elements can be reduced to the calculation of matrix elements of the normal products of field operators (all annihilation operators to the right of the creation operators). In calculating the matrix elements $(\Phi_1^+ S\Phi)$ we shall act with the annihilation operators to the right on the function Φ, and with the creation operators to the left on the function Φ_1^+.

Consider first the result of the action of the operator $A_\mu^{(+)}(x)$ on some state vector Φ. We put the operator $\sum_{\lambda=1}^{4} e_\mu^\lambda(k)\, a_\lambda(k)$ in the form

$$\sum_{\lambda=1}^{4} e_\mu^\lambda(k)\, a_\lambda(k) = \sum_{\lambda=1,2} e_\mu^\lambda(k)\, a_\lambda(k) + e_\mu^4(k)\,(a_4(k)-a_3(k)) + (e_\mu^3(k)+e_\mu^4(k))\, a_3(k). \tag{7.52}$$

From (7.13) it is not difficult to see that

$$e_\mu^3(k)+e_\mu^4(k) = k_\mu/k_0. \tag{7.53}$$

With the help of (7.52) and (7.53) we find that the operator $A_\mu^{(+)}(x)$ is given by

$$A_\mu^{(+)}(x) = A_\mu^{\mathrm{tr}(+)}(x) + \frac{\partial \Lambda^{(+)}(x)}{\partial x_\mu} + L_\mu(x). \tag{7.54}$$

Here

$$\left.\begin{aligned} A_\mu^{\mathrm{tr}(+)}(x) &= \frac{1}{(2\pi)^{3/2}} \int \frac{1}{\sqrt{2k_0}} \sum_{\lambda=1,2} e_\mu^\lambda(k)\, a_\lambda(k)\, e^{ikx}\, dk, \\ \Lambda^{(+)}(x) &= -i\frac{1}{(2\pi)^{3/2}} \int \frac{1}{\sqrt{2k_0}} \frac{1}{k_0} a_3(k)\, e^{ikx}\, dk, \\ L_\mu(x) &= \frac{1}{(2\pi)^{3/2}} \int \frac{1}{\sqrt{2k_0}} e_\mu^4(k)\,(a_4(k)-a_3(k))\, e^{ikx}\, dk. \end{aligned}\right\} \tag{7.55}$$

From (7.45) it follows that

$$L_\mu(x)\Phi = 0. \tag{7.56}$$

As a result of gradient invariance the term $\partial\Lambda^{(+)}/\partial x_\mu$ can be omitted [it is clear from (7.55) that $\Box\Lambda^{(+)}(x) = 0$].

Now let us examine the result of acting with the operator $A_\mu^{(-)}(x)$ on the state vector Φ_1^+. Analogous to (7.54) we obtain

$$A_\mu^{(-)}(x) = A_\mu^{\text{tr}(-)}(x) + \frac{\partial \Lambda^{(-)}(x)}{\partial x_\mu} - L_\mu^+(x), \tag{7.57}$$

where

$$\left.\begin{aligned} A_\mu^{\text{tr}(-)}(x) &= \frac{1}{(2\pi)^{3/2}} \int \frac{1}{\sqrt{2k_0}} \sum_{\lambda=1,2} e_\mu^\lambda(k)\, a_\lambda^+(k)\, \mathrm{e}^{-ikx}\, d\boldsymbol{k}, \\ \Lambda^{(-)}(x) &= \frac{i}{(2\pi)^{3/2}} \int \frac{1}{\sqrt{2k_0}} \frac{1}{k_0} a_3^+(k)\, \mathrm{e}^{-ikx}\, d\boldsymbol{k}. \end{aligned}\right\} \tag{7.58}$$

Using (7.45b) we find

$$\Phi_1^+ L_\mu^+(x) = 0. \tag{7.59}$$

As a result of gradient invariance the derivative $\partial\Lambda^{(-)}/\partial x_\mu$ can be omitted.

It is clear from this discussion that in calculating the matrix elements for the transition between physical states which satisfy the Lorentz conditions (7.45), one is to take into account only A_μ^{tr}, i.e. that part of the operator of the electromagnetic field which provides the creation and annihilation of quanta with transverse polarization.

We turn now to the construction of a set of vectors describing the state of a quantized field. We shall act on the vector Φ_0 with the operators $a_\lambda^+(k)$. From the Lorentz condition (7.45) it follows that the operators $a_3^+(k)$ and $a_4^+(k)$ can appear here only in the combination $a_3^+(k) - a_4^+(k)$. Actually, let us act on the state vector

$$\Phi_1 = (\alpha a_3^+(k) + \beta a_4^+(k))\,\Phi, \tag{7.60}$$

where Φ is a vector satisfying (7.45), with the operator $a_3(k') - a_4(k')$. We obtain

$$\begin{aligned}(a_3(k') - a_4(k'))\,\Phi_1 = \{&[(a_3(k') - a_4(k'))\,(\alpha a_3^+(k) + \beta a_4^+(k))] \\ &+ \alpha(a_3^+(k) + \beta a_4^+(k))\,(a_3(k') - a_4(k'))\}\,\Phi. \end{aligned} \tag{7.61}$$

The second term in the braces when acting on Φ gives zero. From the commutation relations (7.40) it is not difficult to see that the first term is zero, and therefore satisfies the condition (7.45) if $\beta = -\alpha$.

Now let us examine the matrix element $(\Phi_2^+ S \Phi_1)$ where

$$\Phi_1 = (a_3^+(k) - a_4^+(k))\,\Phi$$

[Φ and Φ_2 are arbitrary state vectors satisfying (7.45)].

As we have seen, as a result of gradient invariance and the Lorentz condition the operators $a_\lambda(k)$ and $a_\lambda^+(k)$ can enter the S-matrix only with $\lambda = 1$ and $\lambda = 2$. Thus, using (7.45b) we obtain

$$(\Phi_2^+ S \Phi_1) = \left(\Phi_2^+ (a_3^+(k) - a_4^+(k)) S \Phi\right) = 0. \tag{7.62}$$

It is not difficult to show in an analogous manner that

$$(\Phi_2^+ S \Phi) = 0, \tag{7.63}$$

where

$$\Phi_2 = (a_3^+(k) - a_4^+(k))\,\Phi_1$$

[Φ_1 and Φ are arbitrary state vectors satisfying the conditions (7.45)]. Actually, we have

$$(\Phi_2^+ S\Phi) = \big(\Phi_1^+(a_3(k) - a_4(k))S\Phi\big) = \big(\Phi_1^+ S(a_3(k) - a_4(k))\Phi\big) = 0. \tag{7.64}$$

Thus, as a result of gradient invariance and the Lorentz condition the S-matrix element vanishes if either the initial or the final state vector contains the operators $(a_3^+(k) - a_4^+(k))$.

The vectors which describe the states of the quantized electromagnetic field and which give non-zero transition matrix elements consequently have the form

$$\Phi_{n(k_m,\,\lambda_m)\ldots n(k_1,\,\lambda_1)} = \overset{+}{a}_{\lambda_m}(k_m)\ldots\overset{+}{a}_{\lambda_m}(k_m)\ldots\overset{+}{a}_{\lambda_1}(k_1)\ldots\overset{+}{a}_{\lambda_1}(k_1)\,\Phi_0. \tag{7.65}$$

Here the indices $\lambda_1, \ldots, \lambda_m$ can assume the values 1 and 2. The vector (7.65) describes the state with $n(k_m, \lambda_m)$ quanta with four-momentum k_m and polarization $e^{\lambda_m}(k_m)$, $n(k_{m-1}, \lambda_{m-1})$ quanta with four-momentum k_{m-1} and polarization $e^{\lambda_{m-1}}(k_{m-1})$, and so forth.

CHAPTER 4

EXPANSION OF CHRONOLOGICAL PRODUCTS IN NORMAL PRODUCTS

8. The Interaction Hamiltonians. Normal and Chronological Products of the Field Operators

Our next task is to calculate the matrix elements for the transition between the states which describe free particles with definite momenta. Before proceeding to its solution we shall prove a theorem which enables us to represent the *chronological products* of the field operators in the S-matrix in a form which is extremely convenient for calculating the matrix elements (Wick's theorem).

The S-matrix is defined by the interaction Hamiltonian [see (1.38)]. Let us examine the simplest interaction Hamiltonian. We begin by examining the interaction of the electromagnetic field $A(x)$ with the complex spinor field $\psi(x)$. The electromagnetic potential $A(x)$ must enter the Lagrangian of the system in a form which satisfies gradient invariance. Gradient invariance will be ensured if the electromagnetic potential is introduced by the following replacement in the Lagrangian of the free spinor field [see (3.48)]:

$$\left.\begin{aligned} \frac{\partial\psi}{\partial x_\mu} &\to \left(\frac{\partial}{\partial x_\mu} - ieA_\mu\right)\psi, \\ \frac{\partial\bar{\psi}}{\partial x_\mu} &\to \left(\frac{\partial}{\partial x_\mu} + ieA_\mu\right)\bar{\psi}. \end{aligned}\right\} \tag{8.1}$$

From (6.3) we obtain for the Lagrangian of the field under discussion

$$\mathcal{L} = -\tfrac{1}{2}\left\{\left[\bar{\psi}\gamma_\mu\left(\frac{\partial\psi}{\partial x_\mu} - ieA_\mu\psi\right) + m\bar{\psi}\psi\right] - \left[\left(\frac{\partial\bar{\psi}}{\partial x_\mu} + ieA_\mu\bar{\psi}\right)\gamma_\mu\psi - m\bar{\psi}\psi\right]\right\} + \mathcal{L}_{0e}, \tag{8.2}$$

where $\mathcal{L}_{0e}$ is the Lagrangian of the free electromagnetic field [expression (7.7)]. We write the Lagrangian (8.2) in the form

$$\mathcal{L} = \mathcal{L}_0 + \mathcal{L}_I. \tag{8.3}$$

Here $\mathcal{L}_0$ is the sum of the Lagrangians of the free spinor and electromagnetic fields and $\mathcal{L}_I$ is the interaction Lagrangian. From (8.2) and (8.3) we obtain

$$\mathcal{L}_I(x) = ie\bar{\psi}(x)\gamma_\mu\psi(x)A_\mu(x). \tag{8.4}$$

With the help of the general expression for the current (3.40) and using (8.2) we find the four-vector of current:

$$j_\mu(x) = ie\bar{\psi}(x)\,\gamma_\mu\psi(x). \tag{8.5}$$

Thus, the interaction Lagrangian of the spinor and electromagnetic fields has the form

$$\mathcal{L}_I(x) = j_\mu(x)\,A_\mu(x). \tag{8.6}$$

We have obtained an expression which is well known from classical physics.

We shall now find the Hamiltonian of the system. From the general expression (3.4) we find that the energy density is

$$\mathcal{H} = T_{44} = \frac{\partial\mathcal{L}}{\partial\dfrac{\partial\psi_\sigma}{\partial x_4}}\frac{\partial\psi_\sigma}{\partial x_4} + \frac{\partial\mathcal{L}}{\partial\dfrac{\partial\bar{\psi}_\sigma}{\partial x_4}}\frac{\partial\bar{\psi}_\sigma}{\partial x_4} + \frac{\partial\mathcal{L}}{\partial\dfrac{\partial A_\mu}{\partial x_4}}\frac{\partial A_\mu}{\partial x_4} - \mathcal{L}. \tag{8.7}$$

The interaction Lagrangian (8.4) does not contain any derivatives of the fields and does not contribute to the first three terms of expression (8.7). We obtain

$$\mathcal{H} = \mathcal{H}_0 + \mathcal{L}_I. \tag{8.8}$$

Here $\mathcal{H}_0$ is the sum of the densities of the free field Hamiltonians:

$$\mathcal{H}_0 = \frac{\partial\mathcal{L}_0}{\partial\dfrac{\partial\psi_\sigma}{\partial x_4}}\frac{\partial\psi_\sigma}{\partial x_4} + \frac{\partial\mathcal{L}_0}{\partial\dfrac{\partial\bar{\psi}_\sigma}{\partial x_4}}\frac{\partial\bar{\psi}_\sigma}{\partial x_4} + \frac{\partial\mathcal{L}_0}{\partial\dfrac{\partial A_\mu}{\partial x_4}}\frac{\partial A_\mu}{\partial x_4} - \mathcal{L}_0.$$

From (8.8) it follows that the interaction Hamiltonian of the complex spinor and electromagnetic fields

$$\mathcal{H}_I(x) = -\mathcal{L}_I(x) = -ie\bar{\psi}(x)\,\gamma_\mu\psi(x)\,A_\mu(x). \tag{8.9}$$

Replacing in this expression the functions $\psi(x)$, $\bar{\psi}(x)$, and $A_\mu(x)$ with operators [formulae (6.38) and (7.35)] and acting on the product of field operators with the operator N, we obtain the interaction Hamiltonian in quantum field theory (the interaction Hamiltonian of photons and electrons–positrons). Note that in the units we have chosen $e^2/4\pi \simeq 1/137$.

As our next example let us consider the interaction of the complex spinor field $\psi(x)$ and the real pseudoscalar field $\phi(x)$ (for example, the interaction of neutral π-mesons and nucleons). We construct the interaction Lagrangian by analogy with the Lagrangian (8.4). We assume that the derivatives of the fields do not enter the interaction Lagrangian. Furthermore, the interaction Lagrangian must be scalar. This means that the pseudoscalar $\phi(x)$ must be multiplied by a pseudoscalar constructed from $\psi(x)$ and $\bar{\psi}(x)$. We obtain

$$\mathcal{L}_I(x) = -ig\bar{\psi}(x)\,\gamma_5\psi(x)\,\phi(x). \tag{8.10}$$

It is obvious that this Lagrangian also satisfies the requirements of invariance under gauge transformations. Requiring that $\mathcal{L}_I^* = \mathcal{L}_I$ we find that the constant g is real. The constant g characterizes the interaction between fields (the analog of the electric charge). For the interaction between nucleons and π-mesons $g^2/4\pi \sim 15$. It is not difficult to see that the interaction Hamiltonian of meson and nucleon fields is

$$\mathcal{H}_I(x) = ig\bar{\psi}(x)\,\gamma_5\psi(x)\,\phi(x). \tag{8.11}$$

As our last example we shall consider the Hamiltonian responsible for the β-decay of a neutron:

$$n \to p+e^{-}+\bar{\nu}.$$

Let us assume that there are no derivatives of the field in the interaction Lagrangian. If the weak interaction Lagrangian responsible for the β-decay of the neutron is constructed by analogy to the interaction Lagrangian of the spinor and electromagnetic fields, then the four-vector $(\bar{\psi}_p\gamma_\alpha\psi_n)$ (in quantum theory ψ_p and ψ_n are operators of proton and neutron fields) would have to be multiplied by a four-vector constructed from the operators of electron and neutron fields $(\bar{\psi}_e\gamma_\alpha\psi_\nu)$. Such a Lagrangian would be invariant under space inversion. From experiment it is known that weak interactions do not conserve parity. An effective Lagrangian describing β-decay of the neutron which agrees with the experimental data has the form

$$\mathcal{L}_I = \frac{G}{\sqrt{2}}[(\bar{\psi}_e\gamma_\alpha(1+\gamma_5)\psi_\nu)(\bar{\psi}_p\gamma_\alpha(1+g_A\gamma_5)\psi_n)+(\bar{\psi}_\nu\gamma_\alpha(1+\gamma_5)\psi_e)(\bar{\psi}_n\gamma_\alpha(1+g_A\gamma_5)\psi_p)]. \quad (8.12)$$

The addition of the second term makes the Lagrangian Hermitian. The Lagrangian (8.12) is the sum of a scalar and a pseudoscalar and therefore is not invariant under space inversion. Note that the constants G and g_A are equal to

$$G \approx 10^{-5}\frac{1}{M_p^2}, \quad g_A \simeq 1.2 \quad (8.13)$$

(M_p is the mass of the proton).

We return now to the expression for the S-matrix [see (1.38)]. The interaction Hamiltonian in this expression can be written

$$H_I(x_0) = \int \mathcal{H}_I(x)\, d\mathbf{x}, \quad (8.14)$$

where $\mathcal{H}_I(x)$ is the density of the interaction Hamiltonian ($x_0 \equiv t$). Inserting (8.14) into (1.38) we obtain for the S-matrix

$$S = \sum_{n=0}^{\infty}\frac{(-i)^n}{n!}\int dx_1\int dx_2\ldots\int dx_n P(\mathcal{H}_I(x_1)\ldots\mathcal{H}_I(x_n)), \quad (8.15)$$

where $dx = d\mathbf{x}\, dx_0$.

The interaction Hamiltonian must be given. Our task consists in calculating the matrix elements of the transition (matrix elements of the S-matrix) between states which describe free particles with definite momenta. The operator S is the sum of integrals from the chronological products of field operators. We shall show that the chronological product of field operators can be represented as a sum of normal products (all annihilation operators to the right of every creation operator). The expansion of the chronological products of field operators in normal products allows a significant simplification of the calculation of the S-matrix elements.

We begin by examining the product of two field operators. As an example consider the Hermitian spinor field operators $\psi(x)$ and $\bar{\psi}(x)$ and the operator of a scalar (pseudoscalar)

Hermitian field $\phi(x)$. In the preceding paragraphs we saw that the operators $\psi(x)$, $\bar{\psi}(x)$, and $\phi(x)$ can be written in the forms

$$\left.\begin{aligned} \psi(x) &= \psi^{(+)}(x)+\psi^{(-)}(x),\\ \bar{\psi}(x) &= \bar{\psi}^{(+)}(x)+\bar{\psi}^{(-)}(x),\\ \phi(x) &= \phi^{(+)}(x)+\phi^{(-)}(x). \end{aligned}\right\} \tag{8.16}$$

Here

$$\left.\begin{aligned} \psi^{(+)}(x) &= \frac{1}{(2\pi)^{3/2}}\int\left(\frac{m}{p_0}\right)^{1/2} u^r(p)\, c_r(p)\, e^{ipx}\, d\boldsymbol{p},\\ \bar{\psi}^{(+)}(x) &= \frac{1}{(2\pi)^{3/2}}\int\left(\frac{m}{p_0}\right)^{1/2} \bar{u}^r(-p)\, d_r(p)\, e^{ipx}\, d\boldsymbol{p},\\ \phi^{(+)}(x) &= \frac{1}{(2\pi)^{3/2}}\int\frac{1}{\sqrt{2q_0}}\, a(q)\, e^{iqx}\, d\boldsymbol{q} \end{aligned}\right\} \tag{8.16a}$$

are the positive-frequency parts of operators $\psi(x)$, $\bar{\psi}(x)$, and $\phi(x)$ (the annihilation operators) and

$$\left.\begin{aligned} \psi^{(-)}(x) &= \frac{1}{(2\pi)^{3/2}}\int\left(\frac{m}{p_0}\right)^{1/2} u^r(-p)\, d_r^+(p)\, e^{-ipx}\, d\boldsymbol{p},\\ \bar{\psi}^{(-)}(x) &= \frac{1}{(2\pi)^{3/2}}\int\left(\frac{m}{p_0}\right)^{1/2} \bar{u}^r(p)\, c_r^+(p)\, e^{-ipx}\, d\boldsymbol{p},\\ \phi^{(-)}(x) &= \frac{1}{(2\pi)^{3/2}}\int\frac{1}{\sqrt{2q_0}}\, a^+(q)\, e^{-iqx}\, d\boldsymbol{q} \end{aligned}\right\} \tag{8.16b}$$

are the negative-frequency parts of the corresponding operators (the creation operators). Let us define the normal product operator N. When acting on the product of scalar field operators the operator N arranges them so that all annihilation operators are to the right of the creation operators (normal order). For two operators we have

$$\left.\begin{aligned} N(\phi^{(+)}(x)\,\phi^{(+)}(y)) &= \phi^{(+)}(x)\,\phi^{(+)}(y),\\ N(\phi^{(-)}(x)\,\phi^{(-)}(y)) &= \phi^{(-)}(x)\,\phi^{(-)}(y),\\ N(\phi^{(-)}(x)\,\phi^{(+)}(y)) &= \phi^{(-)}(x)\,\phi^{(+)}(y),\\ N(\phi^{(+)}(x)\,\phi^{(-)}(y)) &= \phi^{(-)}(y)\,\phi^{(+)}(x). \end{aligned}\right\} \tag{8.17}$$

Consider the product $\phi(x)\,\phi(y)$. Using (8.16) and (8.17) we obtain

$$\begin{aligned} \phi(x)\,\phi(y) &= (\phi^{(+)}(x)+\phi^{(-)}(x))(\phi^{(+)}(y)+\phi^{(-)}(y))\\ &= (N(\phi^{(+)}(x)\,\phi^{(+)}(y)) + \phi^{(+)}(x)\,\phi^{(-)}(y)\\ &\quad + N(\phi^{(-)}(x)\,\phi^{(+)}(y)) + N(\phi^{(-)}(x)\,\phi^{(-)}(y)). \end{aligned} \tag{8.18}$$

We further have

$$\phi^{(+)}(x)\,\phi^{(-)}(y) = N(\phi^{(+)}(x)\,\phi^{(-)}(y)) + [\phi^{(+)}(x),\ \phi^{(-)}(y)]. \tag{8.19}$$

From (8.18) and (8.19) we find

$$\phi(x)\,\phi(y) = N(\phi(x)\,\phi(y)) + [\phi^{(+)}(x),\ \phi^{(-)}(y)]. \tag{8.20}$$

Thus, the product of scalar (pseudoscalar) field operators is equal to the normal product of these operators plus the commutator of the positive-frequency part of the first factor and the negative-frequency part of the second factor (the commutator of those operators which in the original product are not arranged in normal order).

When acting on the operators of the spinor field the operator N arranges them in normal order (all annihilation operators to the right of the creation operators). If a rearrangement of field operators occurs, the product of operators arranged in normal order is multiplied by $(-1)^q$, where q is the number of transpositions of spinor field operators which are performed in the transition from the original product of operators to the normal product. For two spinor operators

$$\left.\begin{aligned} N(\psi_\alpha^{(+)}(x)\,\bar\psi_\beta^{(+)}(y)) &= \psi_\alpha^{(+)}(x)\,\bar\psi_\beta^{(+)}(y),\\ N(\psi_\alpha^{(-)}(x)\,\bar\psi_\beta^{(+)}(y)) &= \psi_\alpha^{(-)}(x)\,\psi_\beta^{(+)}(y),\\ N(\psi_\alpha^{(-)}(x)\,\bar\psi_\beta^{(-)}(y)) &= \psi_\alpha^{(-)}(x)\,\bar\psi_\beta^{(-)}(y),\\ N(\psi_\alpha^{(+)}(x)\,\bar\psi_\beta^{(-)}(y)) &= -\bar\psi_\beta^{(-)}(y)\,\psi_\alpha^{(+)}(x), \end{aligned}\right\} \tag{8.21}$$

where α and β are spinor indices.

Note that the sign factor in the definition of the operator N is determined by the commutation relations for spinor field operators. Let us consider, for example, the first of equations (8.21). The operator N acting on the product $\psi_\alpha^{(+)}(x)\,\bar\psi_\beta^{(+)}(y)$ transposes the operators $\psi_\alpha^{(+)}(x)$ and $\bar\psi_\beta^{(+)}(y)$. In accordance with the definition of the operator N we have

$$N(\psi_\alpha^{(+)}(x)\,\bar\psi_\beta^{(+)}(y)) = -\bar\psi_\beta^{(+)}(y)\,\psi_\alpha^{(+)}(x). \tag{8.22}$$

From relations (6.40) and (8.16) it follows that

$$\bar\psi_\beta^{(+)}(y)\,\psi_\alpha^{(+)}(x) = -\psi_\alpha^{(+)}(x)\,\bar\psi_\beta^{(+)}(y),$$

and (8.22) agrees with (8.21). For the scalar field operators we have

$$N(\phi^{(+)}(x)\,\phi^{(+)}(y)) = \phi^{(+)}(y)\,\phi^{(+)}(x),$$

which agrees with (8.17) since the operators $\phi^{(+)}(x)$ and $\phi^{(+)}(y)$ commute.

We now examine the product $\psi_\alpha(x)\,\bar\psi_\beta(y)$:

$$\begin{aligned} \psi_\alpha(x)\,\bar\psi_\beta(y) &= (\psi_\alpha^{(+)}(x)+\psi_\alpha^{(-)}(x))\,(\bar\psi_\beta^{(+)}\psi(y)+\bar\psi_\beta^{(-)}(y))\\ &= N(\psi_\alpha^{(+)}(x)\,\bar\psi_\beta^{(+)}(y))+\psi_\alpha^{(+)}(x)\,\bar\psi_\beta^{(-)}(y)+N(\psi_\alpha^{(-)}(x)\,\bar\psi_\beta^{(+)}(y))+N(\psi_\alpha^{(-)}(x)\,\bar\psi_\beta^{(-)}(y)). \end{aligned} \tag{8.23}$$

Furthermore,

$$\psi_\alpha^{(+)}(x)\,\bar\psi_\beta^{(-)}(y) = N(\psi_\alpha^{(+)}(x)\,\bar\psi_\beta^{(-)}(y))+[\psi_\alpha^{(-)}(x),\,\bar\psi_\beta^{(-)}(y)]_+. \tag{8.24}$$

Inserting (8.24) into (8.23) we find

$$\psi_\alpha(x)\,\bar\psi_\beta(y) = N(\psi_\alpha(x)\,\bar\psi_\beta(y))+[\psi_\alpha^{(+)}(x),\,\bar\psi_\beta^{(-)}(y)]_+. \tag{8.25}$$

From the foregoing it is clear that in the general case

$$U(x)\,V(y) = N(U(x)\,V(y))+[U^{(+)}(x)\,V^{(-)}(y)]_\pm. \tag{8.26}$$

In this expression the minus (commutator) corresponds to the case when $U(x)$ and $V(y)$

are operators of a scalar or pseudoscalar field and the plus (anticommutator) to the case when $U(x)$ and $V(y)$ are spinor field operators. Since

$$[\psi_\alpha^{(+)}(x), \psi_\beta^{(-)}(y)]_+ = 0,$$
$$[\bar{\psi}_\alpha^{(+)}(x), \bar{\psi}_\beta^{(-)}(y)]_+ = 0,$$

then

$$\psi_\alpha(x)\,\psi_\beta(y) = N(\psi_\alpha(x)\,\psi_\beta(y)),$$
$$\bar{\psi}_\alpha(x)\,\bar{\psi}_\beta(y) = N(\bar{\psi}_\alpha(x)\,\bar{\psi}_\beta(y)).$$

Consider the commutator on the right-hand side of relationship (8.20). Using commutation relations (4.26) and expansions (8.16) we obtain

$$\begin{aligned}[\phi^{(+)}(x), \phi^{(-)}(y)] &= \frac{1}{(2\pi)^3}\int \frac{1}{\sqrt{4q_0q_0'}}\, e^{iqx-iq'y}[a(q), a^+(q')]\, dq\, dq' \\ &= \frac{1}{(2\pi)^3}\int \frac{1}{2q_0}\, e^{iq(x-y)}\, dq. \end{aligned} \tag{8.27}$$

The commutator in expression (8.20) is consequently a function of $(x-y)$ and not a field operator.

Furthermore, with the help of (6.40) and (8.16) we find

$$[\psi_\alpha^{(+)}(x), \bar{\psi}_\beta^{(-)}(y)]_+ = \frac{1}{(2\pi)^3}\int \frac{m}{p_0}\, e^{ip(x-y)}\, u_\alpha^r(p)\, \bar{u}_\beta^r(p)\, dp. \tag{8.28}$$

Thus, the normal product operator N is defined so that the product of operators for spinor as well as for scalar fields is equal to the normal product of operators plus a known function (or zero).

Define in the following manner Wick's chronological operator:

$$\left.\begin{aligned} T(\phi(x)\,\phi(y)) &= \begin{cases} \phi(x)\,\phi(y) & x_0 > y_0, \\ \phi(y)\,\phi(x) & y_0 > x_0; \end{cases} \\ T(\psi_\alpha(x)\,\bar{\psi}_\beta(y)) &= \begin{cases} \psi_\alpha(x)\,\bar{\psi}_\beta(y) & x_0 > y_0, \\ -\bar{\psi}_\beta(y)\,\psi_\alpha(x) & y_0 > x_0; \end{cases} \\ T(\psi_\alpha(x)\,\psi_\beta(y)) &= \begin{cases} \psi_\alpha(x)\,\psi_\beta(y) & x_0 > y_0, \\ -\psi_\beta(y)\,\psi_\alpha(x) & y_0 > x_0. \end{cases} \end{aligned}\right\} \tag{8.29}$$

The operator T when acting on the product of scalar field operators arranges them in chronological order (the time argument of the first factor is greater than the time argument of the second) and coincides therefore with the Dyson chronological operator P [see (1.28)]. Acting on the product of spinor field operators the operator T arranges them in chronological order. If there is a transposition of factors, the chronological product of operators is multiplied by -1. In the general case of n factors the action of the operator T results in the field operators being arranged in chronological order (time arguments of the factors decrease from left to right) and the resulting product being multiplied by $(-1)^q$, where q is the number of transpositions of spinor operators which are performed in the transition from the initial product to the chronological product. It is precisely the T-products, as we shall see below, which can be represented in the form of a sum of normal products.

We shall show that in expression (8.15) for the S-matrix the P-products can be replaced by T-products. Let us take as an example the interaction of the electromagnetic and spinor fields. We have

$$T(\mathcal{H}_I(x_1)\,\mathcal{H}_I(x_2)) = \left\{ \begin{array}{ll} \mathcal{H}_I(x_1)\,\mathcal{H}_I(x_2), & x_{10} > x_{20}, \\ \mathcal{H}_I(x_2)\,\mathcal{H}_I(x_1), & x_{20} > x_{10}, \end{array} \right\} = P(\mathcal{H}_I(x_1)\,\mathcal{H}_I(x_2)), \quad (8.30)$$

where $\mathcal{H}_I(x) = -ieN(\bar{\psi}(x)\,\gamma_\mu\psi(x))\,A_\mu(x)$.

In transposing $\mathcal{H}_I(x_1)$ and $\mathcal{H}_I(x_2)$ (for the case $x_{20} > x_{10}$) an even number of transpositions of spinor operators take place. This is related to the fact that the interaction Hamiltonian contains two spinor operators. It is obvious that in the general case of n operators $\mathcal{H}_I$ the T-product and P-product coincide. In all the examples of interactions of field we have discussed there is an even number of spinor operators in the interaction Hamiltonian. From considerations of Lorentz invariance it is clear that this is true for any interaction. Thus, the S-matrix can be written in the form

$$S = \int \sum_{n=0}^{\infty} \frac{(-i)^n}{n!} \int dx_1\, dx_2 \,\ldots\, dx_n\, T(\mathcal{H}_I(x_1)\,\mathcal{H}_I(x_2)\ldots\mathcal{H}_I(x_n)). \quad (8.31)$$

Consider the T-product of two scalar field operators. From (8.29) and (8.32) we obtain

$$T(\phi(x)\,\phi(y)) = \left\{ \begin{array}{ll} N(\phi(x)\,\phi(y)) + [\phi^{(+)}(x), \phi^{(-)}(y)], & x_0 > y_0; \\ N(\phi(y)\,\phi(x)) + [\phi^{(+)}(y), \phi^{(-)}(x)], & y_0 > x_0. \end{array} \right. \quad (8.32)$$

Obviously the scalar field operators can be put under the operator symbol N. We have

$$N(\phi(x)\,\phi(y)) = N(\phi(y)\,\phi(x)). \quad (8.33)$$

From (8.33) and (8.32) we obtain

$$T(\phi(x)\,\phi(y)) = N(\phi(x)\,\phi(y)) + \overline{\phi(x)\,\phi(y)}. \quad (8.34)$$

Here

$$\overline{\phi(x)\,\phi(y)} = \left\{ \begin{array}{ll} [\phi^{(+)}(x), \phi^{(-)}(y)], & x_0 > y_0; \\ [\phi^{(+)}(y), \phi^{(-)}(x)], & y_0 > x_0. \end{array} \right. \quad (8.35)$$

The quantity $\overline{\phi(x)\,\phi(y)}$ is called the *chronological contraction* of the operators $\phi(x)$ and $\phi(y)$. From (8.35) and (8.27) we obtain

$$\overline{\phi(x)\,\phi(y)} = \left\{ \begin{array}{ll} \dfrac{1}{(2\pi)^3} \displaystyle\int \frac{1}{2q_0}\, e^{iq(x-y)}\, dq, & x_0 > y_0; \\ \dfrac{1}{(2\pi)^3} \displaystyle\int \frac{1}{2q_0}\, e^{iq(y-x)}\, dq, & y_0 > x_0. \end{array} \right. \quad (8.36)$$

Thus, the contraction of scalar (pseudoscalar) field operators $\overline{\phi(x)\,\phi(y)}$ is a function of the variable $x-y$. Denote

$$\overline{\phi(x)\,\phi(y)} = \tfrac{1}{2}\Delta_F(x-y; \varkappa). \quad (8.37)$$

Recognizing that $q_0 = \sqrt{\boldsymbol{q}^2+\varkappa^2}$ we write $\frac{1}{2}\Delta_F(x; \varkappa)$ in the form

$$\frac{1}{2}\Delta_F(x;\varkappa) = \begin{cases} \dfrac{1}{(2\pi)^3}\displaystyle\int \dfrac{1}{2\sqrt{\boldsymbol{q}^2+\varkappa^2}} e^{i\boldsymbol{qx}-i\sqrt{\boldsymbol{q}^2+\varkappa^2}\,x_0}\,d\boldsymbol{q}, & x_0 > 0; \\ \dfrac{1}{(2\pi)^3}\displaystyle\int \dfrac{1}{2\sqrt{\boldsymbol{q}^2+\varkappa^2}} e^{-i\boldsymbol{qx}+i\sqrt{\boldsymbol{q}^2+\varkappa^2}\,x_0}\,d\boldsymbol{q}, & x_0 < 0. \end{cases} \tag{8.38}$$

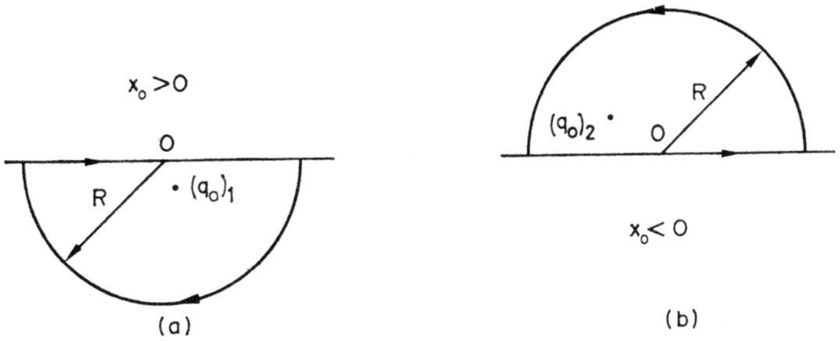

FIG. 2

Let us examine the four-dimensional integral

$$I(x;\varkappa) = -\frac{i}{(2\pi)^4}\lim_{\varepsilon\to 0}\int \frac{e^{iqx}\,dq}{q^2+\varkappa^2-i\varepsilon}, \tag{8.39}$$

where $q = (\boldsymbol{q}, iq_0)$, $dq = d\boldsymbol{q}\,dq_0$ (q_0 is the variable of integration; integration over q_0 goes from $-\infty$ to ∞). We shall show that $\frac{1}{2}\Delta_F(x; \varkappa) = I(x; \varkappa)$. Integrate over the variable q_0 in (8.39). The denominator of the integrated expression vanishes at the points

$$(q_0)_1 = \sqrt{\boldsymbol{q}^2+\varkappa^2}-i\varepsilon', \quad (q_0)_2 = -\sqrt{\boldsymbol{q}^2+\varkappa^2}+i\varepsilon', \tag{8.40}$$

where $\varepsilon' = \varepsilon/2\sqrt{\boldsymbol{q}^2+\varkappa^2}$ (ε is an arbitrarily small quantity; we expanded in powers of ε and kept terms linear in ε). For the case $x_0 > 0$ integration in (8.39) over q_0 from $-\infty$ to ∞ can be replaced by integration over the contour depicted in Fig. 2(a) with $R \to \infty$. The integrated expression in (8.39) has a pole in the lower half-plane at the point $(q_0)_1$. Consequently, the integral (8.39) for $x_0 > 0$ is equal to the residue of the integrated expression at the point $(q_0)_1$ multiplied by $-2\pi i$ (the contour of integration goes clockwise). Then taking the limit $\varepsilon \to 0$ we obtain

$$I(x;\varkappa) = \frac{1}{(2\pi)^3}\int \frac{1}{2\sqrt{\boldsymbol{q}^2+\varkappa^2}} e^{i\boldsymbol{qx}-i\sqrt{\boldsymbol{q}^2+\varkappa^2}\,x_0}\,d\boldsymbol{q}, \qquad x_0 > 0. \tag{8.41a}$$

For $x_0 < 0$ the integral (8.39) is equal to the integral over the contour depicted in Fig. 2(b) with $R \to \infty$. Calculating the residue at the point $(q_0)_2$ we obtain

$$I(x;\varkappa) = \frac{1}{(2\pi)^3}\int \frac{1}{2\sqrt{\boldsymbol{q}^2+\varkappa^2}} e^{i\boldsymbol{qx}+i\sqrt{\boldsymbol{q}^2+\varkappa^2}\,x_0}\,d\boldsymbol{q}, \qquad x_0 < 0. \tag{8.41b}$$

It is easy to see that this expression coincides with the corresponding expression in (8.38).

Thus, the contraction of Hermitian scalar (pseudoscalar) field operators is equal to

$$\overline{\phi(x)\,\phi}(y) = \tfrac{1}{2}\,\Delta_F(x-y;\,\varkappa) = \frac{-i}{(2\pi)^4}\int \frac{e^{iq(x-y)}\,dq}{q^2+\varkappa^2-i\varepsilon}. \tag{8.42}$$

We now examine the T-products of two spinor field operators. From (8.29) and (8.25) we obtain

$$T(\psi_\alpha(x)\,\bar{\psi}_\beta(y)) = \begin{cases} N(\psi_\alpha(x)\,\bar{\psi}_\beta(y)) + [\psi_\alpha^{(+)}(x),\,\bar{\psi}_\beta^{(-)}(y)]_+, & x_0 > y_0; \\ -N(\bar{\psi}_\beta(y)\,\psi_\alpha(x)) - [\bar{\psi}_\beta^{(+)}(y),\,\psi_\alpha^{(-)}(x)]_+, & y_0 > x_0. \end{cases} \tag{8.43}$$

It is not difficult to see that

$$N(\bar{\psi}_\beta(y)\,\psi_\alpha(x)) = -N(\psi_\alpha(x)\,\bar{\psi}_\beta(y)). \tag{8.44}$$

Actually,

$$\left.\begin{aligned} N(\psi_\alpha^{(+)}(x)\,\bar{\psi}_\beta^{(+)}(y)) &= \psi_\alpha^{(+)}(x)\,\bar{\psi}_\beta^{(+)}(y) = -N(\bar{\psi}_\beta^{(+)}(y)\,\psi_\alpha^{(+)}(x)),\\ N(\psi_\alpha^{(+)}(x)\,\bar{\psi}_\beta^{(-)}(y)) &= -\bar{\psi}_\beta^{(-)}(y)\,\psi_\alpha^{(+)}(x) = -N(\bar{\psi}_\beta^{(-)}(y)\,\psi_\alpha^{(+)}(x)),\\ N(\psi_\alpha^{(-)}(x)\,\bar{\psi}_\beta^{(+)}(y)) &= \psi_\alpha^{(-)}(x)\,\bar{\psi}_\beta^{(+)}(y) = -N(\bar{\psi}_\beta^{(+)}(y)\,\psi_\alpha^{(-)}(x)),\\ N(\psi_\alpha^{(-)}(x)\,\bar{\psi}_\beta^{(-)}(y)) &= \psi_\alpha^{(-)}(x)\,\bar{\psi}_\beta^{(-)}(y) = -N(\bar{\psi}_\beta^{(-)}(y)\,\psi_\alpha^{(-)}(x)). \end{aligned}\right\} \tag{8.45}$$

Equation (8.44) follows from these relationships. Inserting (8.44) into (8.43) we obtain

$$T(\psi_\alpha(x)\,\bar{\psi}_\beta(y)) = N(\psi_\alpha(x)\,\bar{\psi}_\beta(y)) + \overline{\psi_\alpha(x)\,\bar{\psi}_\beta}(y), \tag{8.46}$$

where

$$\overline{\psi_\alpha(x)\,\bar{\psi}_\beta}(y) = \begin{cases} [\psi_\alpha^{(+)}(x),\,\bar{\psi}_\beta^{(-)}(y)]_+, & x_0 > y_0; \\ -[\bar{\psi}_\beta^{(-)}(y)\,\psi_\alpha^{(-)}(x)]_+, & y_0 > x_0. \end{cases} \tag{8.47}$$

From (8.28) it is obvious that $\overline{\psi_\alpha(x)\,\bar{\psi}_\beta}(y)$—the chronological contraction of the operators $\psi_\alpha(x)$ and $\bar{\psi}_\beta(y)$—is the α, β element of a matrix function of $x-y$. Note that due to relations (8.44) it is precisely Wick's chronological product which is equal to the normal product plus a function for all values of the variables x and y [the minus sign in (8.44) is compensated for by the corresponding minus sign in the definition of the T-product].

In the general case of two operators $U(x)$ and $V(y)$ we obtain

$$T(U(x)\,V(y)) = N(U(x)\,V(y)) + \overline{U(x)\,V}(y), \tag{8.48}$$

where

$$\overline{U(x)\,V}(y) = \begin{cases} [U^{(+)}(x), \quad V^{(-)}(y)]_\mp, & x_0 > y_0; \\ \pm[V^{(+)}(y), \quad U^{(-)}(x)]_\mp, & y_0 > x_0. \end{cases} \tag{8.49}$$

In this expression the upper signs refer to the case when $U(x)$ and $V(y)$ are scalar (pseudoscalar) field operators and the lower signs refer to spinor fields.

From commutation relations (6.40) and expansions (8.16) it is obvious that

$$\overline{\psi_\alpha(x)\,\psi_\beta}(y) = 0, \quad \overline{\bar{\psi}_\alpha(x)\,\bar{\psi}_\beta}(y) = 0. \tag{8.50}$$

Thus,

$$\left.\begin{aligned} T(\psi_\alpha(x)\,\psi_\beta(y)) &= N(\psi_\alpha(x)\,\psi_\beta(y)),\\ T(\bar{\psi}_\alpha(x)\,\bar{\psi}_\beta(y)) &= N(\bar{\psi}_\alpha(x)\,\bar{\psi}_\beta(y)). \end{aligned}\right\} \tag{8.51}$$

Let us examine the contraction $\overbrace{\psi_\alpha(x)\,\bar{\psi}_\beta(y)}$ defined by relationship (8.47). We have [see (8.28)]

$$
\begin{aligned}
[\psi_\alpha^{(+)}(x),\ \bar{\psi}_\beta^{(-)}(y)]_+ &= \frac{1}{(2\pi)^3}\int e^{ip(x-y)}\,u_\alpha^r(p)\,\bar{u}_\beta^r(p)\,\frac{m}{p_0}\,d\boldsymbol{p}\\
&= \frac{1}{(2\pi)^3}\int e^{ip(x-y)}\,\frac{1}{2p_0 i}\,(\gamma_\mu p_\mu+im)_{\alpha\beta}\,d\boldsymbol{p}\\
&= -i\left(-i\gamma_\mu\frac{\partial}{\partial x_\mu}+im\right)_{\alpha\beta}\frac{1}{(2\pi)^3}\int e^{ip(x-y)}\,\frac{d\boldsymbol{p}}{2p_0}.
\end{aligned}
\tag{8.52}
$$

We used the relationship (see Appendix)

$$
u_\alpha^r(p)\,\bar{u}_\beta^r(p) = \left(\frac{\gamma_\mu p_\mu+im}{2im}\right)_{\alpha\beta}.
$$

From (6.40) and (8.16) we obtain for the anticommutator defining $\overbrace{\psi_\alpha(x)\,\bar{\psi}_\beta(y)}$ for $y_0 > x_0$

$$
\begin{aligned}
[\bar{\psi}_\beta^{(+)}(y),\ \psi_\alpha^{(-)}(x)]_+ &= \frac{-1}{(2\pi)^3}\int e^{-ip(x-y)}\,\bar{u}_\beta^r(-p)\,u_\alpha^r(-p)\,\frac{m}{p_0}\,d\boldsymbol{p}\\
&= \frac{1}{(2\pi)^3}\int e^{ip(y-x)}\,\frac{1}{2ip_0}\,(-\gamma_\mu p_\mu+im)_{\alpha\beta}\,d\boldsymbol{p}\\
&= -i\left(-i\gamma_\mu\frac{\partial}{\partial x_\mu}+im\right)_{\alpha\beta}\int e^{ip(y-x)}\,\frac{1}{2p_0}\,d\boldsymbol{p}.
\end{aligned}
\tag{8.53}
$$

In obtaining this expression we used the relationship

$$
-u_\alpha^r(-p)\,\bar{u}_\beta^r(-p) = \left(\frac{-\gamma_\mu p_\mu+im}{2im}\right)_{\alpha\beta}
$$

(see Appendix). With the help of (8.47), (8.52), and (8.53) we obtain for the contraction of the operators $\psi_\alpha(x)$ and $\bar{\psi}_\beta(y)$ the expression

$$
\overbrace{\psi_\alpha(x)\,\bar{\psi}_\beta(y)} = -i\left(-i\gamma_\mu\frac{\partial}{\partial x_\mu}+im\right)_{\alpha\beta}\tfrac{1}{2}\Delta_F(x-y;\,m),
\tag{8.54}
$$

where

$$
\tfrac{1}{2}\Delta_F(x-y;\,m) = \begin{cases}
\dfrac{1}{(2\pi)^3}\displaystyle\int \frac{e^{ip(x-y)-i\sqrt{m^2+\boldsymbol{p}^2}\,(x_0-y_0)}}{2\sqrt{m^2+\boldsymbol{p}^2}}\,d\boldsymbol{p}, & x_0 > y_0,\\[2ex]
\dfrac{1}{(2\pi)^3}\displaystyle\int \frac{e^{-ip(x-y)+i\sqrt{m^2+\boldsymbol{p}^2}\,(x_0-y_0)}}{2\sqrt{m^2+\boldsymbol{p}^2}}\,d\boldsymbol{p}, & y_0 > x_0.
\end{cases}
$$

This function coincides with (8.38) if one replaces $\varkappa \to m$ in (8.38) and is equal consequently to the four-dimensional integral $I(x-y;\,m)$. We obtain

$$
\overbrace{\psi_\alpha(x)\,\bar{\psi}_\beta(y)} = -i\left(-i\gamma_\mu\frac{\partial}{\partial x_\mu}+im\right)_{\alpha\beta} I(x-y;\,m) = -\frac{1}{(2\pi)^4}\lim_{\varepsilon\to 0}\int\frac{e^{ip(x-y)}\,(\gamma p+im)_{\alpha\beta}}{p^2+m^2-i\varepsilon}\,dp.
\tag{8.55}
$$

Henceforth we shall delete the $-i\varepsilon$ in expressions for chronological contractions [it will always be implicit that the square of the mass is to be replaced by the square of the mass plus $(-i\varepsilon)$]. Expression (8.55) can be written more compactly. Using the commutation relations for the matrices γ_μ we have

$$\frac{(\gamma p - im)(\gamma p + im)}{p^2 + m^2} = 1, \tag{8.56}$$

from which

$$\frac{\gamma p + im}{p^2 + m^2} = \frac{1}{\gamma p - im}. \tag{8.57}$$

The contraction $\overline{\psi(x)\,\bar{\psi}(y)}$ can be written in the form

$$\overline{\psi(x)\,\bar{\psi}(y)} = \frac{-1}{(2\pi)^4} \int \frac{e^{ip(x-y)}}{\gamma p - im}\, dp. \tag{8.58}$$

9. Wick's Theorem

We have examined the products of operators of fields whose quanta are particles with spins 0 and $\frac{1}{2}$. Using general postulates of field theory Pauli showed that operators of fields having particles with any integer spin must satisfy commutation relations of the type (4.26) (quantization with the use of commutators) and operators of fields whose quanta are particles with half odd-integer spin satisfy commutation relations of the type (6.40) (quantization with the use of anticommutators). Particles with integer spin satisfy Bose–Einstein statistics and are called bosons; particles with half odd-integer spin satisfy Fermi–Dirac statistics and are called fermions. The operators of the corresponding fields we shall call, respectively, boson and fermion operators.

We define the normal product operator N in the general case of fields with arbitrary spin. The operator N when acting on the product of creation and annihilation operators arranges them in normal order (all of the annihilation operators to the right of the creation operators). If a permutation of fermion operators takes place, the product of the operators arranged in normal order is multiplied by $\delta = (-1)^q$, where q is the number of transpositions of the fermion operators. From the definition of the operator N it follows that

$$N(UVW\ldots XYZ) = \delta N(UXW\ldots ZVY). \tag{9.1}$$

Here

$$\delta = (-1)^q,$$

where q is the number of transpositions of fermion operators which must be performed in order to go from $UVW\ldots XYZ$ to $UXW\ldots ZVY$. Note that the rule is correct: under the symbol N the field operators can be rearranged since then all boson operators commute while all fermion operators anticommute.

Let us now define in the general case of fields which have particles with arbitrary spin the Wick chronological operator T. When acting on the product of field operators the operator

T arranges them in chronological order (the time argument is left to right). If a permutation of fermion operators takes place, the product of operators arranged in chronological order is multiplied by $\delta = (-1)^q$, where q is the number of transpositions of fermion operators. Obviously (8.48) and (8.49) are true for any field operators [the upper sign in (8.49) refers to boson fields and the lower to fermion fields]. Furthermore, from the definition of the operator T it follows that

$$T(UVW\ldots XYZ) = \delta T(UXW\ldots ZVY). \tag{9.2}$$

Here $\delta_1 = (-1)^q$, where q is the number of transpositions of fermion operators which must be performed to go from $UVW\ldots XYZ$ to $UXW\ldots ZVY$ (the rule for transposing operators under the symbol T is the same as under the symbol N).

Now let us generalize (8.48) to the case of the T-product of an arbitrary number of field operators. Define in the following way the normal product of field operators with contractions of operators under the symbol N:

$$N(\overline{U\underline{VW\ldots R}}\underline{\ldots X}YZ) = \delta \overline{UR}\,\overline{VX}\ldots N(W\ldots YZ). \tag{9.3}$$

Here $\delta = (-1)^q$, where q is the number of transpositions of fermion operators which must be performed to go from $UVW\ldots R\ldots XYZ$ to $URVX\ldots W\ldots YZ$. For example,

$$N(\overline{UVW}R) = -\overline{UW}N(VR),$$

where U, V, W, R are fermion operators.

We shall first prove the following lemma. If the time argument of the operator Z is less than the time argument of the operators $U, V, W, \ldots, X, Y$, then

$$\begin{aligned} N(UVW\ldots XY)Z &= N(UVW\ldots XYZ)+N(UVW\ldots X\overline{YZ})+ \\ &+N(UVW\ldots \overline{XYZ})+\ldots+N(\overline{UVW\ldots XYZ}). \end{aligned} \tag{9.4}$$

It is sufficient to prove relationship (9.4) in the case when $U, V, \ldots, Z$ are creation and annihilation operators.

Let Z be an annihilation operator. It is obvious that the contraction of two annihilation operators is zero. We show that for $x_0 > y_0$

$$\overline{R^{(-)}(x)\,Z^{(+)}(y)} = 0. \tag{9.5}$$

Actually,

$$\begin{aligned} \overline{R^{(-)}(x)\,Z^{(+)}(y)} &= T(R^{(-)}(x)\,Z^{(+)}(y)) - N(R^{(-)}(x)\,Z^{(+)}(y)) \\ &= R^{(-)}(x)\,Z^{(+)}(y) - R^{(-)}(x)\,Z^{(+)}(y) = 0. \end{aligned}$$

Thus, in this case all contractions on the right-hand side of relationship (9.4) are zero and (9.4) reduces to the equation

$$N(UVW\ldots XY)Z = N(UVW\ldots XYZ),$$

which is obvious (Z is an annihilation operator).

Now consider the case with Z a creation operator. Here it is sufficient to prove the lemma for the case when all operators under the symbol N on the left-hand side of (9.4) are annihilation operators. Actually, if relationship (9.4) is proven for this case, then by multiplying its left side by the creation operator D, whose time argument is greater than the time argument of Z, we obtain

$$DN(UVW\ldots XY)Z = N(DUVW\ldots XY)Z = N(DUVW\ldots X\overline{YZ})$$
$$+\ldots+N(D\overline{UVW\ldots XYZ})+N(\overline{DUVW\ldots XYZ}). \tag{9.6}$$

We introduce the operator D under the symbol N, which is always possible since D is a creation operator. Since the contraction of two creation operators is zero, the last added term is zero. If (9.6) is multiplied on the left by a creation operator whose time argument is greater than the time argument of Z, we obtain relationship (9.4) for the case when there are two creation operators under the symbol N on the left side, and so forth. Transposing the operators under the symbol N in both parts of the relationships found in this way, we obtain (9.4) in the general case.

Thus, we proceed to the proof of (9.4) in the only non-trivial case—when Z is a creation operator and $U, V, W, \ldots, X, Y$ are annihilation operators. We have

$$N(UVW\ldots XY)Z = UVW\ldots XYZ. \tag{9.7}$$

Let us examine the product YZ. Since the time argument of operator Y is greater than the time argument of operator Z, $YZ = T(YZ)$. Using (8.48) we obtain

$$YZ = T(YZ) = N(YZ)+\overline{YZ} = \delta_{YZ}ZY+\overline{YZ}, \tag{9.8}$$

where the factor δ_{YZ} is equal to $+1$ (-1) if Y and Z are boson (fermion) operators. Inserting (9.8) into (9.7) and recognizing that

$$UVW\ldots X\overline{YZ} = N(UVW\ldots X\overline{YZ}),$$

we obtain

$$N(UVW\ldots XY)Z = \delta_{YZ}UVW\ldots XZY+N(UVW\ldots X\overline{YZ}). \tag{9.9}$$

Furthermore, replacing the product XZ with the T-product we have

$$XZ = T(XZ) = N(XZ)+\overline{XZ} = \delta_{XZ}ZX+\overline{XZ}. \tag{9.10}$$

It is obvious that

$$\delta_{YZ}UVW\ldots \overline{XZ}Y = N(UVW\ldots \overline{XYZ}). \tag{9.11}$$

From (9.9)–(9.11) we obtain

$$N(UVW\ldots XY)Z = \delta_{XZ}\delta_{YZ}UVW\ldots ZXY$$
$$+N(UVW\ldots X\overline{YZ})+N(UVW\ldots \overline{XYZ}). \tag{9.12}$$

With each application of this procedure the operator Z is moved to the left and an additional term arises which is a normal product which has a contraction of the operator

Z with the operator with which Z is transposed. As a result we obtain

$$N(UVW\ldots XY)\,Z = \delta_{UZ}\delta_{VZ}\ldots\delta_{XZ}\delta_{YZ}ZUVW\ldots XY$$
$$+N(\overline{UVW\ldots XYZ})+\ldots+N(UVW\ldots X\overline{YZ}). \qquad (9.13)$$

From the definition of operator N it follows that

$$\delta_{UZ}\delta_{VZ}\ldots\delta_{XZ}\delta_{YZ}ZUVW\ldots XY = N(UVW\ldots XYZ). \qquad (9.14)$$

Thus, the proof of relationship (9.4) is completed. Relationship (9.4) can be generalized. If (9.4) is multiplied by the contraction $\overline{SQ}$, obviously the resulting relationship can be written in the form

$$N(UV\overline{SW\ldots XQ}Y)Z = N(UV\overline{SW\ldots XQ}YZ)$$
$$+N(UV\overline{SW\ldots XQ}\overline{YZ})+\ldots+N(\underline{\overline{UVSW\ldots XQYZ}}). \quad (9.15)$$

One can multiply (9.4) by several contractions. Thus, relationship (9.4) remains valid with any number of contracted operators under the normal ordering symbol N on each side of the equation.

We now proceed to the proof of the basic theorem—Wick's theorem. We shall show that

$$\begin{aligned} T(UVWR\ldots XYZ) = {} & N(UVWR\ldots XYZ) \\ & +N(\overline{UV}WR\ldots XYZ)+N(\overline{UVW}R\ldots XYZ)+\ldots \\ & +N(\overline{UV}\,\overline{WR}\ldots XYZ)+\ldots+N(\overline{UV}\,\overline{WR\ldots X}YZ)+\ldots . \end{aligned} \qquad (9.16)$$

The right-hand side of this relationship is the sum of normal products with all possible contractions [the second line of (9.16) is the sum of all possible normal products with one contraction, the third line with two, and so forth]. We shall prove (9.16) by induction. Let the theorem be true for the product of n operators $UVWR\ldots XYZ$. Multiplying (9.16) on the right by an operator Ω whose time argument is less than the time arguments of $U, V, W, \ldots, X, Y, Z$ we obtain

$$\begin{aligned} T(UVWR\ldots XYZ)\,\Omega = {} & N(UVWR\ldots XYZ)\,\Omega \\ & +N(\overline{UV}WR\ldots XYZ)\,\Omega+N(\overline{UVW}R\ldots XYZ)\,\Omega+\ldots \\ & +N(\overline{UV}\,\overline{WR}\ldots XYZ)\,\Omega+\ldots . \end{aligned} \qquad (9.17)$$

Applying the lemma proved above to each term on the right-hand side of this equation and recognizing that

$$T(UVWR\ldots XYZ)\,\Omega = T(UVWR\ldots XYZ\Omega),$$

we have verified the correctness of (9.16) for the product of operators $UVWR\ldots XYZ\Omega$ if the time argument of Ω is less than the time arguments of all remaining operators. If on both sides of this relationship the operators are permuted (the rules for permuting field operators under T and N operators are the same), we obtain (9.16) in the general case of

$n+1$ factors. For $n = 2$, relationship (9.16) coincides with (8.48). Thus, relationship (9.16) is proved for the general case.

We now return to the expression for the S-matrix [see (8.31)]. This expression contains the interaction Hamiltonian which, by definition, is the normal product of field operators taken at the same point x.

Thus, the S-matrix is the sum of integrals of

$$T(N(UVW)\ldots N(XYZ)), \tag{9.18}$$

where the field operators under any N symbol have the same time argument. Such T-products are called mixed products.

In order to apply expansion (9.16) to the mixed T-products we proceed as follows. We add the arbitrarily small positive quantity ε to the time arguments of all creation operators in the mixed T-product. Then the operators N in (9.18) can be deleted. Actually, consider the example $N(U^{(+)}(x_0)\,V^{(+)}(x_0)\,W^{(-)}(x_0))$ (the variable $\boldsymbol{x}$ on which these operators depend does not concern us here). We obtain

$$N(U^{(+)}(x_0)\,V^{(+)}(x_0)\,W^{(-)}(x_0)) = \delta W^{(-)}(x_0)\,U^{(+)}(x_0)\,V^{(+)}(x_0), \tag{9.19}$$

where $\delta = \pm 1$ depending on the parity of the number of fermion operator transpositions. On the other hand

$$T(U^{(+)}(x_0)\,V^{(+)}(x_0)\,W^{(-)}(x_0+\varepsilon)) = \delta W^{(-)}(x_0+\varepsilon)\,U^{(+)}(x_0)\,V^{(+)}(x_0) \tag{9.20}$$

[the operators on the right-hand side of (9.20) are arranged in the same order as in (9.19); the factor δ is the same].

Thus, if ε is added to the time argument of all creation operators in the mixed T-product, the mixed T-product can be replaced by the usual T-product and represented in accordance with (9.16) in the form of a sum of normal products with all possible contractions. Then ε is to be put equal to zero.

It is clear that the contractions of operators which enter the mixed T-product under the same N symbol are zero. If the time argument of a creation operator is greater than the time argument of an annihilation operator, the contraction of these operators is zero [see (9.5)]. The contractions of creation operators and the contractions of annihilation operators are also zero. Thus, expansion (9.16) is correct for the mixed T-product provided one omits the contractions of operators which are under the same operator symbol N and which have the same space-time argument.

In the following paragraphs we shall see how much Wick's theorem simplifies the calculation of S-matrix elements. Before turning to the examination of concrete processes we shall discuss various terms in expansion (9.16). We shall verify that many of them are equivalent. As an example consider the interaction of the electromagnetic field $A(x)$ and the electron–positron field $\psi(x)$. The interaction Hamiltonian is given by the expression

$$\mathscr{H}_I(x) = -ieN(\bar{\psi}(x)\,\gamma_\mu(\psi x))\,A_\mu(x).$$

Let us examine the first term in the series (8.31). Using Wick's theorem for the mixed

T-product we obtain

$$T\big(N(\bar\psi(x)\,\gamma_\mu\psi(x))\,A_\mu(x)\big) = N(\bar\psi(x)\,\gamma_\mu\psi(x))\,A_\mu(x). \tag{9.21}$$

Let us now look at the second term of expansion (8.31). Since the operators ψ and A commute (different fields),

$$\begin{aligned}&T\big(N(\bar\psi(x_1)\,\gamma_\mu\psi(x_1))\,A_\mu(x_1)\,N(\bar\psi(x_2)\,\gamma_\nu\psi(x_2))\,A_\nu(x_2)\big)\\ &\quad= T\big(N(\bar\psi(x_1)\,\gamma_\mu\psi(x_1))\,N(\bar\psi(x_2)\,\gamma_\nu\psi(x_2))\big)\,T(A_\mu(x_1)\,A_\nu(x_2)).\end{aligned} \tag{9.22}$$

Using Wick's theorem for mixed T-products and taking into account that $\overbracket{\bar\psi(x_1)\,\bar\psi(x_2)} = 0$ and $\overbracket{\psi(x_1)\,\psi(x_2)} = 0$, we obtain

$$\begin{aligned}&T\big(N(\bar\psi(x_1)\,\gamma_\mu\psi(x_1))\,N(\bar\psi(x_2)\,\gamma_\nu\psi(x_2))\big) = N(\bar\psi(x_1)\,\gamma_\mu\psi(x_1)\,\bar\psi(x_2)\,\gamma_\nu\psi(x_2))\\ &\quad+N(\overbracket{\bar\psi(x_1)\gamma_\mu\psi(x_1)\bar\psi(x_2)\,\gamma_\nu\psi(x_2)})+N(\bar\psi(x_1)\,\gamma_\mu\overbracket{\psi(x_1)\,\bar\psi(x_2)}\,\gamma_\nu\psi(x_2))\\ &\quad+N(\overbracket{\bar\psi(x_1)\,\gamma_\mu\overbracket{\psi(x_1)\,\bar\psi(x_2)}\,\gamma_\nu\psi(x_2)}).\end{aligned} \tag{9.23}$$

The second term on the right-hand side of expansion (9.23) contributes the following to the S-matrix:

$$\int N(\overbracket{\bar\psi(x_1)\,\gamma_\mu\psi(x_1)\,\bar\psi(x_2)\,\gamma_\nu\psi(x_2)})\,T(A_\mu(x_1)\,A_\nu(x_2))\,dx_1\,dx_2. \tag{9.24}$$

Transposing under the N symbol the operators

$$\bar\psi(x_2)\,\gamma_\nu\psi(x_2) \qquad \text{and} \qquad \bar\psi(x_1)\,\gamma_\mu\psi(x_1)$$

and interchanging the variables of integration ($x_1 \rightleftarrows x_2$) we find

$$\begin{aligned}&\int N(\overbracket{\bar\psi(x_1)\gamma_\mu\psi(x_1)\bar\psi(x_2)\gamma_\nu\psi(x_2)})\,T(A_\mu(x_1)\,A_\nu(x_2))\,dx_1\,dx_2\\ &\quad= \int N(\bar\psi(x_1)\,\gamma_\nu\overbracket{\psi(x_1)\,\bar\psi(x_2)}\,\gamma_\mu\psi(x_2))\,T(A_\mu(x_2)\,A_\nu(x_1))\,dx_2\,dx_1\\ &\quad= \int N(\bar\psi(x_1)\,\gamma_\mu\overbracket{\psi(x_1)\,\bar\psi(x_2)}\,\gamma_\nu\psi(x_2))\,T(A_\mu(x_1)\,A_\nu(x_2))\,dx_1\,dx_2.\end{aligned} \tag{9.25}$$

To obtain the last equation we used the fact that

$$T(A_\mu(x_2)\,A_\nu(x_1)) = T(A_\nu(x_1)\,A_\mu(x_2)),$$

and reassigned the summation indices μ and ν ($\mu \rightleftarrows \nu$).

It is obvious from (9.25) that the second and third terms in expansion (9.23) make identical contributions to the S-matrix. Thus, these terms are equivalent. One needs to take only one of them into account and delete the 2! in the denominator of the corresponding term in expansion (8.31).

Let us now examine the T-product

$$T\big(N(\bar\psi(x_1)\,\gamma_\mu\psi(x_1))\,N(\bar\psi(x_2)\,\gamma_\nu\psi(x_2))\,N(\bar\psi(x_3)\,\gamma_\varrho\psi(x_3))\big) \tag{9.26}$$

found in the next term of the series (8.31). It is not difficult to see that in the expansion of

(9.26) into normal products there are 3! normal products with one contraction. By means of the corresponding permutations of field operators and relabeling of the variables of integration and summation, it is easy to verify the equivalence of these terms. Furthermore, it is not difficult to see that in the expansion of the T-product (9.26) there are 3! equivalent normal products with two contractions in which the uncontracted operators have different arguments, e.g.

$$N(\bar{\psi}(x_1)\, \gamma_\mu \overbrace{\psi(x_1)\, \bar{\psi}(x_2)}\, \gamma_\nu \overbrace{\psi(x_2)\, \bar{\psi}(x_3)}\, \gamma_\varrho \psi(x_3))$$

and so forth, and three equivalent normal products with two contractions in which the uncontracted operators have the same argument, e.g.

$$N(\overbrace{\bar{\psi}(x_1)\, \gamma_\mu \overbrace{\psi(x_1)\, \bar{\psi}(x_2)}\, \gamma_\nu \psi(x_2)}\, \bar{\psi}(x_3)\, \gamma_\varrho \psi(x_3))$$

and so forth. Finally, the expansion contains two equivalent normal products with three contractions.

In the general case of an arbitrary normal product, all normal products equivalent to it can be obtained by transposing pairs of operators $\bar{\psi}\gamma\psi$ under the N symbol and relabeling the variables correspondingly. The number of such equivalent normal products which make identical contributions to the S-matrix is equal to $n!/g$, where n is the number of pairs $\bar{\psi}\gamma\psi$ under the symbol of the operator N and g is the number of transpositions which do not change the form of the normal product (after the corresponding relabeling of variables).

CHAPTER 5

RULES FOR CONSTRUCTING FEYNMAN DIAGRAMS

10. Electron Scattering by an External Electromagnetic Field. Feynman Diagrams

We now use the apparatus we have developed to study particle collision. We begin with the problem of electron scattering by an unquantized external electromagnetic field (for example, electron scattering by the Coulomb field of a nucleus).

The interaction Hamiltonian has the form

$$\mathscr{H}_I(x) = -ieN(\bar{\psi}(x)\,\gamma_\mu\psi(x))\,A_\mu(x), \tag{10.1}$$

where $\bar{\psi}(x)$ and $\psi(x)$ are electron–positron field operators and the potential $A_\mu(x)$ is a given function (not a field operator).

Let the electron at the initial time ($t \to -\infty$) have four-momentum p and let the spin projection of the electron in the direction of the momentum be r. We shall find the probability amplitude for detection of an electron at a later time ($t \to \infty$) with four-momentum p' and spin projection r' in the direction of the momentum.

The initial and final state vectors are respectively

$$\Phi_{pr} = c_r^+(p)\Phi_0, \quad \Phi_{p'r'} = c_{r'}^+(p')\Phi_0. \tag{10.2}$$

We must calculate the matrix element for the transition between the states Φ_{pr} and $\Phi_{p'r'}$, i.e.

$$(\Phi_{p'r'}^+ S\Phi_{pr}) = (\Phi_0^+ c_{r'}(p')\,Sc_r^+(p)\Phi_0). \tag{10.3}$$

We write the S-matrix in the form

$$S = \sum_{n=0}^{\infty} S^{(n)}, \tag{10.4}$$

where

$$S^{(n)} = \frac{(-i)^n}{n!}\int dx_1 \ldots dx_n\, T(\mathscr{H}_I(x_1)\ldots\mathscr{H}_I(x_n)).$$

The operator $S^{(n)}$ is proportional to e^n. Since the constant e is small ($e^2/4\pi \simeq 1/137$), perturbation theory is applicable and it is sufficient to consider only the contribution to the transition matrix element of terms of lowest order in e.

We shall calculate the matrix elements of the first few operators in expansion (10.4) and formulate the rules (*Feynman rules*) which allow one to obtain easily the matrix element of any operator $S^{(n)}$.

In calculating the matrix elements the chronological products will be represented according to Wick's theorem as a sum of normal products with all possible contractions.

The operator N arranges the field operators so that the annihilation operators are located to the right of the creation operators. In calculating the transition matrix elements we propose to act on the initial state vector with the operators $\psi^{(+)}$ and $\bar{\psi}^{(+)}$ and on the final state vector with the operators $\psi^{(-)}$ and $\bar{\psi}^{(-)}$.

Let us examine the first-order term

$$T(\bar{\psi}(x)\,\gamma_\mu\psi(x)) = N(\bar{\psi}(x)\,\gamma_\mu\psi(x)) = \bar{\psi}^{(+)}(x)\,\gamma_\mu\psi^{+}(x)$$
$$+N(\bar{\psi}^{(+)}(x)\,\gamma_\mu\psi^{(-)}(x))+\bar{\psi}^{(-)}(x)\,\gamma_\mu\psi^{(+)}(x)+\bar{\psi}^{(-)}(x)\,\gamma_\mu\psi^{(-)}(x). \tag{10.5}$$

Furthermore,

$$N(\bar{\psi}^{(+)}(x)\,\gamma_\mu\psi^{(-)}(x)) = N(\bar{\psi}_\alpha^{(-)}(x)\,(\gamma_\mu)_{\alpha\beta}\psi_\beta^{(-)}(x))$$
$$= -\psi_\beta^{(-)}(x)(\gamma_\mu)_{\alpha\beta}\bar{\psi}_\alpha^{(+)}(x) = -\psi^{(-)}(x)\gamma_\mu\bar{\psi}^{(+)}(x). \tag{10.6}$$

We act with the operator $\psi_\beta^{(+)}(x)$ on the initial state vector Φ_{pr}:

$$\psi_\beta^{(+)}(x)\,c_r^+(p)\,\Phi_0 = ([\psi_\beta^{(+)}(x),\,c_r^+(p)]_+ - c_r^+(p)\,\psi_\beta^{(+)}(x))\,\Phi_0. \tag{10.7}$$

Using commutation relations (6.40) and expansion (8.16) we find

$$[\psi_\beta^{(+)}(x),\,c_r^+(p)]_+ = \frac{1}{(2\pi)^{3/2}}\int\left(\frac{m}{p_0'}\right)^{1/2} u_\beta^{r'}(p')\,[c_{r'}(p'),\,c_r^+(p)]_+\,e^{ip'x}\,dp'$$
$$= \frac{1}{(2\pi)^{3/2}}\left(\frac{m}{p_0}\right)^{1/2} u_\beta^r(p)\,e^{ipx}. \tag{10.8}$$

It is obvious that $\psi_\beta^{(+)}(x)\Phi_0 = 0$. We ultimately obtain

$$\psi_\beta^{(+)}(x)\,c_r^+(p)\,\Phi_0 = \frac{1}{(2\pi)^{3/2}}\left(\frac{m}{p_0}\right)^{1/2} u_\beta^r(p)\,e^{ipx}\,\Phi_0. \tag{10.9}$$

Thus, by acting with the operator $\psi^{(+)}(x)$ on the state vector which describes the electron with four-momentum p and spin projection r we obtain the vector Φ_0 which is multiplied by the function which is the corresponding solution of the free Dirac equation [the operator $\psi^{(+)}(x)$ annihilates the electron]. Note that in the general case if Φ is the vector describing any state in which there are no electrons,

$$\psi_\beta^{(+)}(x)\,c_r^+(p)\,\Phi = ([\psi_\beta^{(+)}(x),\,c_r^+(p)]_+ - c_r^+(p)\,\psi_\beta^{(+)}(x))\,\Phi = \frac{1}{(2\pi)^{3/2}}\left(\frac{m}{p_0}\right)^{1/2} u_\beta^r(p)\,e^{ipx}\,\Phi. \tag{10.10}$$

In obtaining this relationship we used relationship (10.8) and the condition

$$\psi_\beta^{(+)}(x)\,\Phi = 0. \tag{10.11}$$

In order to calculate the matrix element of the operator $\bar{\psi}^{(+)}(x)\,\gamma_\mu\psi^{(+)}(x)$, we act with the operator $\bar{\psi}_\alpha^{(+)}(x)\,\psi_\beta^{(+)}(x)$ on the initial state vector. Since $\bar{\psi}_\alpha^{(+)}(x)\,\Phi_0 = 0$, we find from (10.9)

$$\bar{\psi}_\alpha^{(+)}(x)\,\psi_\beta^{(+)}(x)\,c_r^+(p)\,\Phi_0 = 0$$

[the operator $\psi^{(+)}(x)$ annihilates the electron; the second annihilation operator acting on the vacuum vector gives zero].

One can verify this in another way. Using commutation relations (6.40) we obtain

$$\bar{\psi}_\alpha^{(+)}(x)\,\psi_\beta^{(+)}(x)\;c_r^+(p)\Phi_0 = \psi_\beta^{(+)}(x)\;c_r^+(p)\,\bar{\psi}_\alpha^{(+)}(x)\,\Phi_0 = 0.$$

Obviously in general if there are only particles and no antiparticles in the initial state, the operator $\bar{\psi}^{(+)}$ acting on the initial state vector gives zero. Thus, the operator $\bar{\psi}^{(+)}(x)\,\gamma_\mu\psi^{(+)}(x)$ does not contribute to the matrix element we are calculating. It is also clear that the contribution of the operator $\psi^{(-)}(x)\,\tilde{\gamma}_\mu\bar{\psi}^{(+)}(x)$ to this matrix element is zero.

Let us examine the two remaining operators on the right-hand side of expansion (10.5). We act with the operator $\bar{\psi}_\alpha^{(-)}(x)$ to the left on the vector $\Phi^+_{p'r'}$. We have

$$\Phi_0^+c_{r'}(p')\,\bar{\psi}_\alpha^{(-)}(x) = \Phi_0^+([c_{r'}(p'),\bar{\psi}_\alpha^{(-)}(x)]_+ - \bar{\psi}_\alpha^{(-)}(x)\,c_{r'}(p')). \tag{10.12}$$

With the help of commutation relations (6.40) we find

$$\begin{aligned}[c_{r'}(p')\,\bar{\psi}_\alpha^{(-)}(x)]_+ &= \frac{1}{(2\pi)^{3/2}}\int\left(\frac{m}{p_0'}\right)^{1/2}\bar{u}_\alpha^r(p)\,\mathrm{e}^{-ipx}\,[c_{r'}(p'),c_r^+(p)]_+\,d\boldsymbol{p}\\ &= \frac{1}{(2\pi)^{3/2}}\left(\frac{m}{p_0'}\right)^{1/2}\mathrm{e}^{-ip'x}\,\bar{u}_\alpha^{r'}(p').\end{aligned} \tag{10.13}$$

We obtain from the condition (6.46) by Hermitian conjugation

$$\Phi_0^+c_r^+(p) = 0,\quad \Phi_0^+d_r^+(p) = 0.$$

From this we find that

$$\Phi_0^+\bar{\psi}_\alpha^{(-)}(x) = 0,\quad \Phi_0^+\psi_\beta^{(-)}(x) = 0. \tag{10.14}$$

From (10.12)–(10.14) we obtain

$$\Phi_0^+c_{r'}(p')\,\bar{\psi}_\alpha^{(-)}(x) = \Phi_0^+\frac{1}{(2\pi)^{3/2}}\left(\frac{m}{p_0'}\right)^{1/2}\bar{u}_\alpha^{r'}(p')\,\mathrm{e}^{-ip'x}. \tag{10.15}$$

By acting with the operator $\psi^{(-)}(x)$ to the left on the final vector $\Phi_0^+c_{r'}(p')$ we obtain the vector Φ_0^+ which is multiplied by the function which is the solution of the free Dirac equation for the conjugate spinor. In the general case if the vector Φ_1 satisfies the condition

$$\Phi_1^+\psi_\alpha^{(-)}(x) = 0$$

(Φ_1 is the vector describing the state in which there are no electrons),

$$\begin{aligned}\Phi_1^+c_{r'}(p')\,\bar{\psi}_\alpha^{(-)}(x) &= \Phi_1^+([c_{r'}(p'),\bar{\psi}_\alpha^{(-)}(x)]_+ - \bar{\psi}_\alpha^{(-)}(x)\,c_{r'}(p'))\\ &= \Phi_1^+\frac{1}{(2\pi)^{3/2}}\left(\frac{m}{p_0'}\right)^{1/2}\bar{u}_\alpha^{r'}(p')\,\mathrm{e}^{-ip'x}.\end{aligned} \tag{10.16}$$

In this manner by acting to the left the operator $\bar{\psi}^{(-)}(x)$ annihilates the electron. This can be shown in another way. From (8.16) it is not difficult to see that

$$\bar{\psi}_\alpha^{(-)}(x) = (\psi_\beta^{(+)}(x))^+\,(\gamma_4)_{\beta\alpha}.$$

From this we find

$$\Phi^+_{p'r'}\bar{\psi}^{(-)}_\alpha(x) = (\psi^{(+)}_\beta(x)\,\Phi_{p'r'})^+(\gamma_4)_{\beta\alpha}. \tag{10.17}$$

From (10.17) and (10.9) we obtain (10.15).

Let us now consider the result of acting with the operator $\bar{\psi}^{(-)}_\alpha(x)\,\psi^{(-)}_\beta(x)$ to the left on the final state. With the help of (10.15) and (10.14) we obtain

$$\Phi^+_0 c_{r'}(p')\,\bar{\psi}^{(-)}_\alpha(x)\,\psi^{(-)}_\beta(x) = 0$$

[the operator $\bar{\psi}^{(-)}$ in acting to the left on $\Phi^+_0 c_{r'}(p')$ annihilates the electron and the second operator $\psi^{(-)}(x)$ gives zero when acting to the left on Φ^+_0]. This is easy to show in another manner. Transposing the operators we find

$$\Phi^+_0 c_{r'}(p')\,\bar{\psi}^{(-)}_\alpha(x)\,\psi^{(-)}_\beta(x) = \Phi^+_0\psi^{(-)}_\beta(x)\,c_{r'}(p')\,\bar{\psi}^{(-)}_\alpha(x) = 0. \tag{10.18}$$

We obtain zero since there are no positrons in the state $\Phi_{p'r'}$. It is obvious that in the general case the operator $\psi^{(-)}$ acting to the left will give zero if there are no positrons in the final state. Thus, the operator $\bar{\psi}^{(-)}(x)\,\gamma_\mu\psi^{(-)}(x)$ also does not contribute to the matrix elements we are considering.

The operator $\bar{\psi}^{(-)}(x)\,\gamma_\mu\psi^{(+)}(x)$ makes a non-zero contribution. Using (10.9) and (10.15) we obtain

$$\begin{aligned}(\Phi^+_{p'r'}S^{(1)}\Phi_{pr}) &= (-i)\int(\Phi^+_0 c_{r'}(p')\,\bar{\psi}^{(-)}_\alpha(x)\,(\gamma_\mu)_{\alpha\beta}(-ie)\,A_\mu(x)\,\psi^{(+)}_\beta(x)\,c_r(p)\,\Phi_0)\,dx\\ &= (-i)\,\Phi^+_0\left[\int\frac{1}{(2\pi)^{3/2}}\left(\frac{m}{p'_0}\right)^{1/2}\bar{u}^{r'}_\alpha(p')\,\mathrm{e}^{-ip'x}(-ie)\,(\gamma_\mu)_{\alpha\beta}A_\mu(x)\right.\\ &\quad\left.\times u^r_\beta(p)\,\mathrm{e}^{ipx}\,\frac{1}{(2\pi)^{3/2}}\left(\frac{m}{p_0}\right)^{1/2}dx\right]\Phi_0.\end{aligned} \tag{10.19}$$

It is obvious that the integral in this expression is not a field operator. Recognizing that the vector Φ_0 is normalized ($\Phi^+_0\Phi_0 = 1$) we obtain

$$\begin{aligned}(\Phi^+_{p'r'}S^{(1)}\Phi_{pr}) &= (-i)\,\frac{1}{(2\pi)^{3/2}}\left(\frac{m}{p'_0}\right)^{1/2}\bar{u}^{r'}(p')\,(-ie)\,\gamma_\mu u^r(p)\,\frac{1}{(2\pi)^{3/2}}\\ &\quad\times\left(\frac{m}{p_0}\right)^{1/2}A_\mu(p'-p).\end{aligned} \tag{10.20}$$

Here

$$A_\mu(p'-p) = \int \mathrm{e}^{-i(p'-p)x}\,A_\mu(x)\,dx. \tag{10.21}$$

We now calculate the matrix element of the operator $S^{(2)}$. In the preceding section the mixed T-product in $S^{(2)}$ was expanded in normal products [see (9.23)]. Consider first the operator

$$N(\bar{\psi}(x_1)\,\gamma_\mu\psi(x_1)\,\bar{\psi}(x_2)\,\gamma_\nu\psi(x_2)).$$

Obviously it does not contribute to the matrix element we are interested in. Actually, let us put the operators ψ and $\bar{\psi}$ in the form of a sum of positive-frequency and negative-frequency parts. We obtain a sum of products each of which contains four field operators. It is clear that each product contains not less than two positive-frequency or negative-frequency operators. Two positive-frequency operators acting on the initial vector describing one electron give zero. Two negative-frequency operators acting to the left on a final state vector describing one electron also give zero.

Let us examine the normal products with one contraction. In expansion (9.23) there are two such terms which make, as we have shown, the same contribution to $S^{(2)}$. Therefore it is sufficient to examine one of them, e.g.

$$N(\bar{\psi}(x_1)\,\gamma_\mu \overbrace{\psi(x_1)\,\bar{\psi}(x_2)}\,\gamma_\nu\psi(x_2)). \tag{10.22}$$

If $\psi(x_2)$ and $\bar{\psi}(x_1)$ are expanded into positive-frequency and negative-frequency components, we obtain four terms. From the above it is clear that in calculating the matrix element under consideration, one needs to take into account only one of them, namely

$$\bar{\psi}^{(-)}_\alpha(x_1)\,(\gamma_\mu)_{\alpha\beta}\,\overbrace{\psi_\beta(x_1)\,\bar{\psi}_\varrho(x_2)}\,(\gamma_\nu)_{\varrho\sigma}\,\psi^{(+)}_\sigma(x_2). \tag{10.23}$$

Using (10.9), (10.15), and (8.58) we obtain

$$\begin{aligned}(\Phi^+_{p'r'}S^{(2)}\,\Phi_{pr}) &= (-i)^2\Phi_0^+\int\frac{1}{(2\pi)^{3/2}}\left(\frac{m}{p_0'}\right)^{1/2} \mathrm{e}^{-ip'x_1}\bar{u}^{r'}_\alpha(p')\,(-ie)\,(\gamma_\mu)_{\alpha\beta}\\ &\quad\times\frac{(-1)}{(2\pi)^4}\left[\left[\int dp_1\left(\frac{1}{\gamma p_1-im}\right)_{\beta\varrho}\mathrm{e}^{ip_1(x_1-x_2)}\right](-ie)\,(\gamma_\nu)_{\varrho\sigma}\,\mathrm{e}^{ipx_2}\,u^r_\sigma(p)\right.\\ &\quad\times\left(\frac{m}{p_0}\right)^{1/2}\frac{1}{(2\pi)^{3/2}}A_\mu(x_1)\,A_\nu(x_2)\,dx_1\,dx_2\,\Phi_0\\ &= (-i)^2\int\frac{1}{(2\pi)^{3/2}}\left(\frac{m}{p_0'}\right)^{1/2}\bar{u}^{r'}(p')\,(-ie)\,\gamma_\mu\frac{(-1)}{(2\pi)^4}\frac{1}{\gamma p_1-im}\\ &\quad\times(-ie)\,\gamma_\nu u^r(p)\frac{1}{(2\pi)^{3/2}}\left(\frac{m}{p_0}\right)^{1/2}A_\mu(p'-p_1)\,A_\nu(p_1-p)\,dp_1.\end{aligned} \tag{10.24}$$

We now examine the matrix element of the operator $S^{(3)}$. A non-zero contribution to the matrix element of the process of electron scattering by an external field is made by the operator

$$\bar{\psi}^{(-)}(x_1)\,\gamma_\mu\overbrace{\psi(x_1)\,\bar{\psi}(x_2)}\,\gamma_\nu\overbrace{\psi(x_2)\,\bar{\psi}(x_3)}\,\gamma_\varrho\psi^{(+)}(x_3). \tag{10.25}$$

In the expansion of the chronological product in $S^{(3)}$ there are 3! terms in the normal product equivalent to (10.25). Using (10.8), (10.15), and (8.58) we obtain

$$\begin{aligned}(\Phi_0^+c_{r'}(p')\,S^{(3)}\,c_r^+(p)\,\Phi_0) &= (-i)^3\int\frac{1}{(2\pi)^{3/2}}\left(\frac{m}{p_0'}\right)^{1/2}\bar{u}^{r'}(p')\,(-ie)\,\gamma_\mu\frac{(-1)}{(2\pi)^4}\\ &\quad\times\frac{1}{\gamma p_1-im}(-ie)\,\gamma_\nu\frac{(-1)}{(2\pi)^4}\frac{1}{\gamma p_2-im}(-ie)\,\gamma_\varrho u^r(p)\frac{1}{(2\pi)^{3/2}}\\ &\quad\times\left(\frac{m}{p_0}\right)^{1/2}A_\mu(p'-p_1)\,A_\nu(p_1-p_2)\,A_\varrho(p_2-p)\,dp_1\,dp_2.\end{aligned} \tag{10.26}$$

Now it is easy to write the contribution of any operator $S^{(n)}$ to the matrix element of the process under consideration. The terms of any order in the expansion of the matrix element in e are constructed from the same quantities arranged in a definite order and having the same factors everywhere: the initial spinor $u^r(p)$, the γ matrices, the Fourier components of the electromagnetic potential, the operators $1/(\gamma p - im)$, and the final spinor $\bar{u}^{r'}(p')$. This is determined by the fact that a contribution to the matrix element under consideration is made only by those operators which are constructed from the operator $\psi^{(+)}$ which annihilates the initial electron, the γ matrices, the potential A, the contractions, and the operator $\psi^{(-)}$ which annihilates the final electron.

If one puts the quantities in the matrix element in correspondence to the lines and vertices of a diagram, the diagrams with terms of any order in the expansion of the matrix element in e will be constructed from the same elements. We shall put the first-order matrix element [see (10.20)] in correspondence with the diagram shown in Fig. 3.

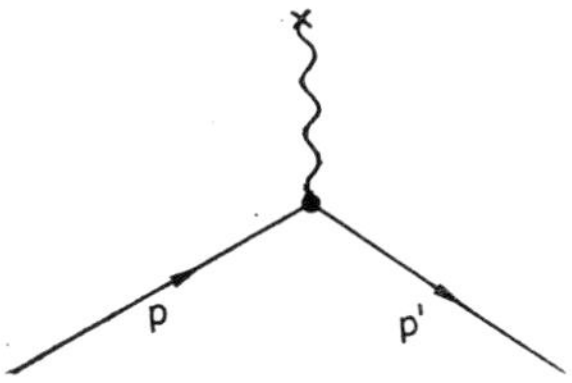

FIG. 3

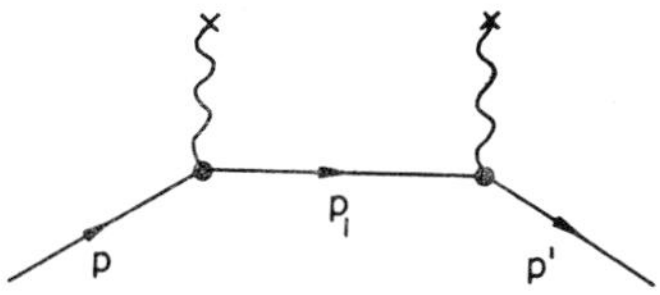

FIG. 4

Let the straight line emerging from the diagram marked with momentum p' correspond in the matrix element to the spinor

$$\frac{1}{(2\pi)^{3/2}}\left(\frac{m}{p_0'}\right)^{1/2} \bar{u}^{r'}(p'),$$

the entering line with momentum p to the spinor

$$\frac{1}{(2\pi)^{3/2}}\left(\frac{m}{p_0}\right)^{1/2} u^{r}(p),$$

the vertex to the matrix $(-ie)\gamma_\mu$, and the wavy line with the cross at the end to the Fourier component of the electromagnetic potential $A_\mu(p'-p)$ (the argument is the difference between the emerging and entering four-momenta).

To obtain the matrix element (10.20) it is necessary, moving in the direction opposite the arrows along the straight lines, to write out the quantities on the straight lines and the vertex in the diagram in Fig. 3, and then multiply that expression by the Fourier component of the potential corresponding to the wavy line. The resulting expression is then multiplied by $-i$.

We put the matrix element (10.24) (the next-order term) in correspondence with the diagram in Fig. 4. In comparison with the diagram in Fig. 3, this diagram has one new element—an internal line. We mark this line with the momentum p_1 and require that it correspond to the operator

$$-\frac{1}{(2\pi)^4}\frac{1}{\gamma p_1 - im}.$$

We assume the same rules of correspondence for the external lines and vertices of the diagram in Fig. 4 as for the diagram in Fig. 3. It is obvious that to obtain the matrix element (10.24) it is necessary, moving opposite the arrows along the solid lines, to write out in sequence those quantities to which the lines and vertices correspond, to multiply the resulting expression by the Fourier components of the potential to which the wavy lines correspond, and then to integrate that expression over the momentum p_1. The expression found thereby is then multiplied by $(-i)^2$.

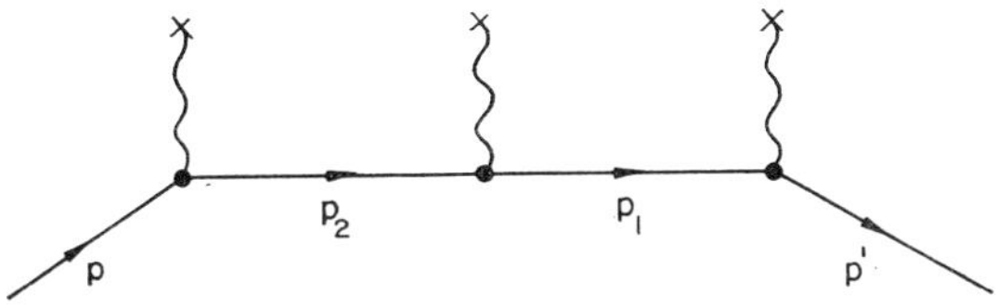

FIG. 5

If one accepts the rules formulated above, it is not difficult to see that the diagram in Fig. 5 corresponds to the third-order matrix element (10.26).

The diagram for the term of nth order obviously consists of n vertices and n wavy lines (the number of interaction Hamiltonians), $n-1$ internal lines (the number of contractions), and two solid external lines (unpaired operators).

Thus, the diagrams provide a graphical representation of the terms of the perturbation theory series. They are called *Feynman diagrams*. With the help of Feynman diagrams it is easy to account for (by perturbation theory) all possibilities determined by the interaction Hamiltonian for the transitions of a system from the initial to the final state.

In Fig. 3 the electron interacts with the external field and goes from the state with momentum p to the state with momentum p'. In Fig. 4 the electron interacts with the external field and goes from the state with momentum p to the intermediate (virtual) state with momentum p_1 ($p_1^2 \neq m^2$, p_1 is the integration variable), is propagated in this state, then again interacts with the external field and goes into the final state with momentum p. The operator

$$\frac{(-1)}{(2\pi)^4} \frac{1}{\gamma p_1 - im},$$

to which the internal line of the diagram corresponds (representing the propagation of the electron in the virtual state), is called the electron *propagator*. In Fig. 5 the electron goes from the initial state to the intermediate state with momentum p_2, then goes to the intermediate state with momentum p_1, and finally goes from there to the final state.

Diagrams of higher orders differ from these diagrams by additional transitions to intermediate states caused by the interaction with the external field. An electron interacting with an external field can go from a real state to a real state, from a real to a virtual (from a virtual to a real), or from a virtual to a virtual state. These possibilities are accounted for by the Feynman diagrams.

In the following sections we shall examine the interaction of quantized electromagnetic and electron–positron fields and shall learn how to construct Feynman diagrams for various processes in which electron, positrons, and photons take part.

To conclude let us return to the examination of the operator $S^{(3)}$. In the expansion in $S^{(3)}$ of the mixed T-product in normal products, in addition to the normal product with two contractions which we have already considered there is the term

$$N(\overline{\psi}(x_1)\,\gamma_\mu\psi(x_1)\,\overline{\psi}(x_2)\,\gamma_\nu\psi(x_2)\,\overline{\psi}(x_3)\,\gamma_\varrho\psi(x_3)). \tag{10.27}$$

This operator is the product of the operator $N(\overline{\psi}(x_3)\,\gamma_\varrho\,\psi(x_3))$ and the function $\overline{\psi}(x_1)\,\gamma_\mu\,\psi(x_1)\,\overline{\psi}(x_2)\,\gamma_\nu\,\psi(x_2)$ which does not have any variables in common with the operator. The matrix element of the operator (10.27) is proportional consequently to the matrix element of the operator $S^{(1)}$.

The normal products which separate into the product of two factors with no variables in common (the field operator and the product of contractions) are not to be considered in calculating the transition matrix element. Let us group the terms of the S-matrix proportional to $N(\overline{\psi}\gamma_\mu\psi)$. Taking into account the equivalence of terms differing only by the transposition of variables we obtain

$$(-i)\int(-ie)\,N(\overline{\psi}(x)\,\gamma_\mu\psi(x))\,A_\mu(x)\,dx\left[1+\frac{(-i)^2}{2!}(-ie)^2\right.$$
$$\left.\times\int N(\overline{\psi}(x_1)\,\gamma_\mu\psi(x_1)\overline{\psi}(x_2)\gamma_\nu\psi(x_2)\,A_\mu(x_1))\,A_\nu(x_2)\,dx_1\,dx_2+\dots\right]. \tag{10.28}$$

It is not difficult to see that the series in square brackets is the matrix element

$$(\Phi_0^+ S\Phi_0). \tag{10.29}$$

Obviously only those terms of the S-matrix in which all field operators are contracted contribute to the matrix element (10.29). We have

$$(\Phi_0^+ S^{(0)}\Phi_0) = 1, \quad (\Phi_0^+ S^{(1)}\Phi_0) = 0,$$
$$(\Phi_0^+ S^{(2)}\Phi_0) = \frac{(-i)^2}{2!}\int(-ie)^2\,\overline{\psi}(x_1)\,\gamma_\mu\psi(x_1)\,\overline{\psi}(x_2)\,\gamma_\nu\psi(x_2)\times A_\mu(x_1)\,A_\nu(x_2)\,dx_1\,dx_2$$

and so forth.

Similarly, one can verify that the matrix element of any term of the S-matrix which is not separated into the product of factors with no variables in common (a field operator and contractions) is multiplied by $(\Phi_0^+\,S\Phi_0)$.

The matrix element for the process of electron scattering by an external field consequently can be written in the form

$$(\Phi_{p'r'}^+ S\Phi_{pr}) = (\Phi_0^+ S\Phi_0)\,R_{p'r';pr}, \tag{10.30}$$

where the matrix element $R_{p'r';pr}$ corresponds to the diagrams in Figs. 3–5, and so forth.

The state vectors (10.2) describe an electron and a vacuum. We are interested in the probability amplitude for the transition of the electron from the state with momentum p

into the state with momentum p' with the condition that the vacuum remain a vacuum (a conditional probability). In order to obtain this quantity it is necessary to divide the full amplitude of the transition $(\Phi^+_{p'r'}\, S\Phi_{pr})$ by the probability amplitude that the vacuum remains a vacuum, i.e. by $(\Phi_0^+ S\Phi_0)$. Thus, the matrix element of electron scattering is given by the relationship

$$\frac{(\Phi^+_{p'r'} S\Phi_{pr})}{(\Phi_0^+ S\Phi_0)}\,. \tag{10.31}$$

Note that this remark relates not only to the process of electron scattering by an external field, but to all physical processes.

11. Processes in which Electrons, Positrons, and Photons take part

SCATTERING OF A PHOTON BY AN ELECTRON

In this section we shall examine the interaction of quantized electromagnetic and electron–positron fields and formulate the rules for constructing Feynman diagrams for the processes in which electrons, positrons, and photons take part.

We begin with the examination of the process of photon scattering by an electron (Compton effect):

$$\gamma + e \to \gamma + e. \tag{11.1}$$

Let k and p (k' and p') be respectively the four-momenta of the initial (final) photon and electron, r and r' the spin projections of the initial and final electron, and λ and λ' the polarization vectors of the initial and final photons (the initial and final photons can have only transverse polarization, i.e. λ and λ' take the values 1 and 2). The initial and final state vectors are equal to:

$$\left.\begin{aligned} \Phi_{pr,\,k\lambda} &= c_r^+(p)\, a_\lambda^+(k)\, \Phi_0, \\ \Phi_{p'r',\,k'\lambda'} &= c_{r'}^+(p')\, a_{\lambda'}^+(k')\, \Phi_0. \end{aligned}\right\} \tag{11.2}$$

Let us now look at the matrix element

$$(\Phi^+_{p'r',\,k'\lambda'} S\Phi_{pr,\,k\lambda}). \tag{11.3}$$

The operator S is given by expression (10.4) and the interaction Hamiltonian is

$$\mathcal{H}_I(x) = -ieN(\bar{\psi}(x)\, \gamma_\mu \psi(x))\, A_\mu(x), \tag{11.4}$$

where $\psi(x)$, $\bar{\psi}(x)$, and $A_\mu(x)$ are field operators.

We are going to act on the initial and final state vectors first with the operators of the electron–positron field and then with the operators of the electromagnetic field. We shall show that the operator $S^{(1)}$ does not contribute to the matrix element of the process (11.1). It is obvious that of the four terms arising in the separation of $\psi(x)$ and $\bar{\psi}(x)$ into positive-frequency and negative-frequency parts, only the operator $\bar{\psi}^{(-)}(x)\, \gamma_\mu \psi^{(+)}(x)$ makes a non-zero contribution to the matrix element. From (10.10) and (10.16) we obtain

$$\left.\begin{aligned} \psi_\beta^{(+)}(x)\, c_r^+(p)\, a_\lambda^+(k)\Phi_0 &= \frac{1}{(2\pi)^{3/2}}\left(\frac{m}{p_0}\right)^{1/2} u_\beta^r(p)\, e^{ipx}\, a_\lambda^+(k)\,\Phi_0, \\ \Phi_0^+ a_{\lambda'}(k')\, c_{r'}(p')\, \bar{\psi}_\alpha^{(-)}(x) &= \Phi_0^+ a_{\lambda'}(k')\frac{1}{(2\pi)^{3/2}}\left(\frac{m}{p_0'}\right)^{1/2} \bar{u}_\alpha^{r'}(p')\, e^{-ip'x}. \end{aligned}\right\} \tag{11.5}$$

In order to find the matrix element $S^{(1)}$, it is necessary consequently to calculate†

$$(\Phi_0^+ a_{\lambda'}(k')\, A_\mu^{\text{tr}}(x)\, a_\lambda^+(k)\,\Phi_0). \tag{11.6}$$

The operator $A_\mu^{\text{tr}}(x)$ is equal to

$$A_\mu^{\text{tr}}(x) = A_\mu^{\text{tr}(+)}(x) + A_\mu^{\text{tr}(-)}(x), \tag{11.7}$$

where

$$\left.\begin{aligned} A_\mu^{\text{tr}(+)}(x) &= \frac{1}{(2\pi)^{3/2}}\int \frac{1}{\sqrt{2k_0}}\sum_{\lambda=1,2} e_\mu^\lambda(k)\, a_\lambda(k)\, e^{ikx}\, d\mathbf{k}, \\ A_\mu^{\text{tr}(-)}(x) &= \frac{1}{(2\pi)^{3/2}}\int \frac{1}{\sqrt{2k_0}}\sum_{\lambda=1,2} e_\mu^\lambda(k)\, a_\lambda^+(k)\, e^{-ikx}\, d\mathbf{k}. \end{aligned}\right\} \tag{11.8}$$

We act with the operator $A_\mu^{\text{tr}(+)}(x)$ to the right on the vector $a_\lambda^+(k)\Phi_0$:

$$A_\mu^{\text{tr}(+)}(x)\, a_\lambda^+(k)\,\Phi_0 = \left([A_\mu^{\text{tr}(+)}(x),\, a_\lambda^+(k)] + a_\lambda^+(k)\, A_\mu^{\text{tr}(+)}(x)\right)\Phi_0. \tag{11.9}$$

Using commutation relations (7.40) we obtain from (11.8)

$$\begin{aligned} [A_\mu^{\text{tr}(+)}(x), a_\lambda^+(k)] &= \frac{1}{(2\pi)^{3/2}}\int \frac{1}{\sqrt{2k_0'}}\sum_{\lambda'=1,2} e_\mu^{\lambda'}(k')\, e^{ik'x}[a_{\lambda'}(k'),\, a_\lambda^+(k)]\, d\mathbf{k}' \\ &= \frac{1}{(2\pi)^{3/2}}\frac{1}{\sqrt{2k_0}}\, e_\mu^\lambda(k)\, e^{ikx}. \end{aligned} \tag{11.10}$$

Since

$$A_\mu^{\text{tr}(+)}(x)\,\Phi_0 = 0, \tag{11.11}$$

then from (11.9) and (11.10) we find

$$A_\mu^{\text{tr}(+)}(x)\, a_\lambda^+(k)\,\Phi_0 = \frac{1}{(2\pi)^{3/2}}\frac{1}{\sqrt{2k_0}}\, e_\mu^\lambda(k)\, e^{ikx}\,\Phi_0. \tag{11.12}$$

Note that if Φ is any vector satisfying the condition

$$A_\mu^{\text{tr}(+)}(x)\,\Phi = 0 \tag{11.13}$$

(Φ describes the state in which there are no photons with transverse polarization), then in an analogous manner

$$A_\mu^{\text{tr}(+)}(x)\, a_\lambda^+(k)\Phi = \frac{1}{(2\pi)^{3/2}}\frac{1}{\sqrt{2k_0}}\, e_\mu^\lambda(k)\, e^{ikx}\,\Phi. \tag{11.14}$$

† As was shown in § 7, due to the Lorentz condition and gradient invariance it is necessary to consider only the transverse part of the unpaired electromagnetic field operators in calculating the matrix elements.

We act with the operator $A_\mu^{\text{tr}(-)}(x)$ to the left on the vector $\Phi_0^+ a_{\lambda'}(k')$. We find

$$\Phi_0^+ a_{\lambda'}(k')\, A_\mu^{\text{tr}(-)}(x) = \Phi_0^+([a_{\lambda'}(k'),\, A_\mu^{\text{tr}(-)}(x)] + A_\mu^{\text{tr}(-)}(x)\, a_{\lambda'}(k'))$$
$$= \Phi_0^+ \frac{1}{(2\pi)^{3/2}} \frac{1}{\sqrt{2k_0'}} e_\mu^{\lambda'}(k')\, \mathrm{e}^{-ik'x}. \tag{11.15}$$

In obtaining this relationship we used commutation relations (7.40), expansion (11.8), and the condition

$$\Phi_0^+ A_\mu^{\text{tr}(-)}(x) = 0, \tag{11.16}$$

which follows from

$$\Phi_0^+ a_\lambda^+(k) = (a_\lambda(k)\Phi_0)^+ = 0, \qquad \lambda = 1, 2.$$

Note that the relationship (11.15) can be obtained in another way. From (11.8) it is obvious that

$$A_\mu^{\text{tr}(-)}(x) = (A_\mu^{\text{tr}(+)}(x))^+. \tag{11.17}$$

We find

$$\Phi_0^+ a_{\lambda'}(k')\, A_\mu^{\text{tr}(-)}(x) = (A_\mu^{\text{tr}(+)}(x)\, a_{\lambda'}^+(k')\, \Phi_0)^+. \tag{11.18}$$

Using (11.12) we obtain (11.15) from this.

If the vector Φ_1 describes any state in which there are no transverse photons ($\Phi_1^+ A_\mu^{\text{tr}(-)}(x) = 0$), then obviously

$$\Phi_1^+ a_{\lambda'}(k')\, A_\mu^{\text{tr}(-)}(x) = \Phi_1^+ \frac{1}{(2\pi)^{3/2}} \frac{1}{\sqrt{2k_0'}} e_\mu^{\lambda'}(k')\, \mathrm{e}^{-ik'x}. \tag{11.19}$$

We now examine the matrix element (11.6). Using (11.7), (11.12), and (11.15) we obtain

$$(\Phi_0^+ a_{\lambda'}(k')\, A_\mu^{\text{tr}}(x)\, a_\lambda^+(k)\, \Phi_0) = \frac{1}{(2\pi)^{3/2}} \left[\frac{1}{\sqrt{2k_0}} e_\mu^\lambda(k)\, \mathrm{e}^{ikx} (\Phi_0^+ a_{\lambda'}(k')\, \Phi_0) \right.$$
$$\left. + \frac{1}{\sqrt{2k_0'}} e_\mu^{\lambda'}(k')\, \mathrm{e}^{-ik'x} (\Phi_0^+ a_\lambda^+(k)\, \Phi_0) \right]. \tag{11.20}$$

It is obvious that

$$(\Phi_0^+ a_{\lambda'}(k')\, \Phi_0) = 0, \quad (\Phi_0^+ a_\lambda^+(k)\, \Phi_0) = 0. \tag{11.21}$$

Thus,

$$(\Phi_{p'r',\, k'\lambda'} S^{(1)} \Phi_{pr,\, k\lambda}) = 0.$$

We now turn to an examination of the matrix element of the operator $S^{(2)}$. It is clear that in the expansion of the T-product of electron–positron field operators into normal products a non-zero contribution is made by the term

$$\overline{\psi}^{(-)}(x_1)\, \gamma_\mu \underbrace{\psi(x_1)\, \overline{\psi}(x_2)}\, \gamma_\nu \psi^{(+)}(x_2) \tag{11.22}$$

[the operator $\psi^{(+)}(x_2)$ annihilates the initial electron and $\bar{\psi}^{(-)}(x_1)$ annihilates the final electron]. Remember that there are 2! such terms in the expression for the S-matrix. Using (11.5) we obtain:

$$(\Phi^+_{p'r',\,k'\lambda'}S^{(2)}\Phi_{pr,\,k\lambda}) = (-i)^2 \int dx_1\,dx_2\,dp_1 \Bigg[\frac{1}{(2\pi)^{3/2}} \left(\frac{m}{p_0'}\right)^{1/2}$$
$$\times \bar{u}^{r'}(p')\,\mathrm{e}^{-ip'x_1}(-ie)\,\gamma_\mu \frac{(-1)}{(2\pi)^4}\,\mathrm{e}^{ip_1(x_1-x_2)} \frac{1}{\hat{p}_1 - im}(-ie)$$
$$\times \gamma_\nu u^r(p)\,\mathrm{e}^{ipx_2} \frac{1}{(2\pi)^{3/2}} \left(\frac{m}{p_0}\right)^{1/2} \left(\Phi_0^+ a_{\lambda'}(k')\,T(A_\mu(x_1)\,A_\nu(x_2))a^+_\lambda(k)\,\Phi_0\right)\Bigg], \tag{11.23}$$

where $\hat{p}_1 = p_1\gamma = p_{1\mu}\gamma_\mu$. We shall use this notation extensively in the future.

In accordance with Wick's theorem we have

$$T(A_\mu(x_1)\,A_\nu(x_2)) = N(A_\mu(x_1)\,A_\nu(x_2)) + \overline{A_\mu(x_1)\,A_\nu}(x_2). \tag{11.24}$$

We further obtain†

$$N(A_\mu(x_1)\,A_\nu(x_2)) = A_\mu^{(+)}(x_1)\,A_\nu^{(+)}(x_2) + A_\mu^{(-)}(x_1)\,A_\nu^{(+)}(x_2) + A_\nu^{(-)}(x_2)\,A_\mu^{(+)}(x_1) + A_\mu^{(-)}(x_1)\,A_\nu^{(-)}(x_2). \tag{11.25}$$

It is obvious that the operator $A_\mu^{(+)}(x_1)\,A_\nu^{(+)}(x_2)$ does not contribute to the matrix element (11.23) [the operator $A_\nu^{(+)}(x_2)$ annihilates the initial photon, then the operator $A_\mu^{(+)}(x_1)$ acting on the vector Φ_0 gives zero]. The contribution of the operator $A_\mu^{(-)}(x_1)\,A_\nu^{(-)}(x_2)$ is also zero. A non-zero contribution to the matrix element (11.23) is made only by the second and third terms of the expansion (11.25). With the help of (11.14) and (11.19) we find

$$\left(\Phi_0^+ a_{\lambda'}(k')\,N(A_\mu(x_1)\,A_\nu(x_2))\,a^+_\lambda(k)\,\Phi_0\right)$$
$$= (\Phi_0^+\Phi_0)\Bigg[\frac{1}{(2\pi)^{3/2}}\frac{1}{\sqrt{2k_0'}}e_\nu^{\lambda'}(k')\,\mathrm{e}^{-ik'x_2}\frac{1}{(2\pi)^{3/2}}\frac{1}{\sqrt{2k_0}}e_\mu^{\lambda}(k)\,\mathrm{e}^{ikx_1}$$
$$+\frac{1}{(2\pi)^{3/2}}\frac{1}{\sqrt{2k_0'}}e_\mu^{\lambda'}(k')\,\mathrm{e}^{-ik'x_1}\frac{1}{(2\pi)^{3/2}}\frac{1}{\sqrt{2k_0}}e_\nu^{\lambda}(k)\,\mathrm{e}^{ikx_2}\Bigg]. \tag{11.26}$$

We insert (11.26) into (11.23). Integrating over the variables x_1 and x_2 and using the normalization condition $(\Phi_0^+\Phi_0) = 1$, we obtain‡

† In calculating the matrix elements of the S-matrix one must consider only the transverse part of the operator A, i.e. A^{tr} (see § 7). Henceforth the index tr on uncontracted electromagnetic field operators will be omitted.

‡ In order to establish the rules of correspondence between the matrix elements and diagrams, we write out all functions and operators together with those factors which arise with them.

$$(\Phi_{p'r',\,k'\lambda'}S^{(2)}\Phi_{pr,\,k\lambda}) = (-i)^2\left\{\int\left[\frac{1}{(2\pi)^{3/2}}\left(\frac{m}{p_0'}\right)^{1/2}\bar{u}^{r'}(p')(-ie)\,\gamma_\mu(2\pi)^4\right.\right.$$

$$\times\,\delta(p'+k'-p_1)\frac{(-1)}{(2\pi)^4}\frac{1}{\hat{p}_1-im}(-ie)\,\gamma_\nu(2\pi)^4\,\delta(p_1-p-k)\,u^r(p)$$

$$\left.+\frac{1}{(2\pi)^{3/2}}\left(\frac{m}{p_0}\right)^{1/2}\frac{1}{(2\pi)^{3/2}}\frac{1}{\sqrt{2k_0'}}e_\mu^{\lambda'}(k')\frac{1}{(2\pi)^{3/2}}\frac{1}{\sqrt{2k_0}}e_\nu^{\lambda}(k)\right]dp_1$$

$$+\int\left[\frac{1}{(2\pi)^{3/2}}\left(\frac{m}{p_0'}\right)^{1/2}\bar{u}^{r'}(p')(-ie)\,\gamma_\mu(2\pi)^4\,\delta(p'-k-p_1)\right.$$

$$\times\frac{(-1)}{(2\pi)^4}\frac{1}{\hat{p}_1-im}(-ie)\,\gamma_\nu(2\pi)^4\,\delta(p_1+k'-p)\,u^r(p)\frac{1}{(2\pi)^{3/2}}$$

$$\left.\left.\times\left(\frac{m}{p_0}\right)^{1/2}\frac{1}{(2\pi)^{3/2}}\frac{1}{\sqrt{2k_0}}e_\mu^{\lambda}(k)\frac{1}{(2\pi)^{3/2}}\frac{1}{\sqrt{2k_0'}}e_\nu^{\lambda'}(k')\right]dp_1\right\}. \qquad (11.27)$$

We put the first term in (11.27) in correspondence with the Feynman diagram in Fig. 6. The straight emerging and entering lines correspond to the final and initial spinors

$$\frac{1}{(2\pi)^{3/2}}\left(\frac{m}{p_0'}\right)^{1/2}\bar{u}(p') \quad\text{and}\quad \frac{1}{(2\pi)^{3/2}}\left(\frac{m}{p_0}\right)^{1/2}u(p).$$

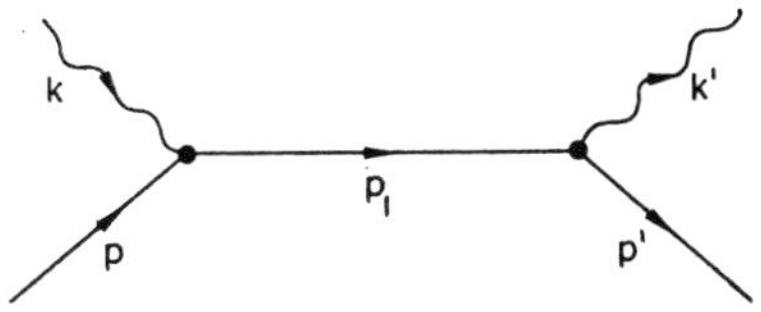

FIG. 6

The internal line marked with the momentum p corresponds to the matrix

$$\frac{(-1)}{(2\pi)^4}\frac{1}{\hat{p}_i-im}.$$

The first vertex corresponds to $(-ie)\,\gamma_\nu(2\pi)^4\,\delta(p_1-p-k)$ and the wavy line entering this vertex corresponds to the polarization vector of the initial photon $e_\nu^\lambda(k)$ multiplied by $(1/(2\pi)^{3/2})\,(1/\sqrt{2k_0}.)$ The argument of the δ-function is the difference between the four-momentum emerging from the vertex and the sum of the entering four-momenta. Thus, the δ-function ensures the conservation of the total four-momentum at the vertex.

Let the second vertex in the diagram correspond to the matrix

$$(-ie)\,\gamma_\mu(2\pi)^4\,\delta(p'+k'-p_1),$$

and the wavy line emerging from this vertex correspond to the factor

$$\frac{1}{(2\pi)^{3/2}}\frac{1}{\sqrt{2k_0'}}e_\mu^{\lambda'}(k')$$

(the argument of the δ-function is the difference between the total emerging and total entering four-momenta).

If one accepts these rules, then as can be seen from Fig. 6 and expression (11.27) to obtain the matrix element it is necessary, moving along the solid line opposite the arrows (beginning at the end), to write out all elements to which the straight lines and vertices correspond. The resulting expression is multiplied by the quantities to which the wavy lines correspond, summed over the indices μ and ν, and integrated over the momentum of the internal line p_1.

The diagram in Fig. 6 describes the following sequence of interaction events. The initial electron absorbs the initial photon and goes into the intermediate (virtual) state with momentum p_1. Then the electron propagates in the intermediate state. Finally, the virtual electron with momentum p_1 emits the final photon and goes into the final state. All momenta of the virtual electron are integrated over.

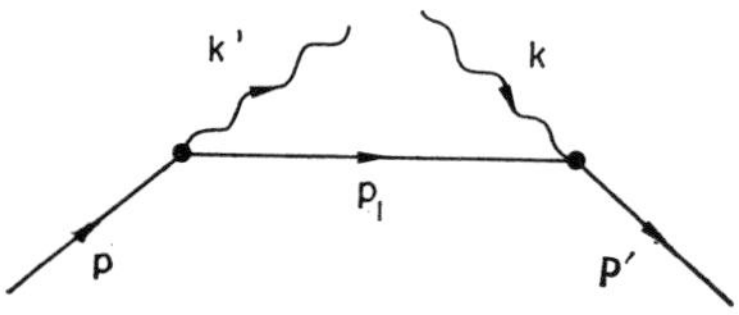

FIG. 7

The diagram vertices are determined by the interaction Hamiltonian. A solid line (ψ and $\bar{\psi}$ in the Hamiltonian) enters and emerges from each vertex; in addition, a wavy line (operator A) either enters or emerges from the vertex. The conservation of the total four-momentum at each vertex ensures the conservation of the total four-momentum in the scattering process. Note that the number of vertices coincides with the order of e in perturbation theory.

It is obvious that besides this possibility (described by the diagram in Fig. 6) for the transition from the initial into the final state there is another, namely: the initial electron can first emit a final photon, go into the intermediate state, and then absorb the initial photon and go into the final state. This possibility is depicted in the diagram in Fig. 7.

In accordance with the formulated rules, we must put

$$(-ie)\,\gamma_\nu(2\pi)^4\,\delta(p_1+k'-p)$$

into correspondence with the first vertex and

$$(-ie)\,\gamma_\mu(2\pi)^4\,\delta(p'-k-p_1)$$

with the second vertex (the argument of the first as well as of the second δ-function is the difference between the total emerging and entering four-momenta). The wavy line emerging from the first vertex is to be put into correspondence with

$$\frac{1}{(2\pi)^{3/2}}\,\frac{1}{\sqrt{2k_0'}}\,e_\nu^{\lambda'}(k'),$$

and the wavy line entering the second vertex with

$$\frac{1}{(2\pi)^{3/2}} \frac{1}{\sqrt{2k_0}} e_\mu^\lambda(k).$$

Moving opposite the arrows along the solid line and writing out the terms belonging to the elements of the diagram, we obtain the second term of the matrix element (11.27).

Thus, in the second order of perturbation theory the process of photon–electron scattering is described by two diagrams.

Obviously it is possible to integrate over the momentum of the intermediate electron. As a result we obtain the following expression for the matrix element of the Compton effect in the second order of perturbation theory:

$$(\Phi^+_{p'r',\,k'\lambda'} S^{(2)} \Phi_{pr,\,k\lambda}) = (-i)^2 \frac{1}{(2\pi)^6} \left(\frac{m^2}{4p_0 p_0' k_0 k_0'} \right)^{1/2} \frac{(-1)}{(2\pi)^4} (-ie)^2 (2\pi)^8 \, \delta(p'+k'-p-k)$$
$$\times \left[\bar{u}^{r'}(p')\, e_\mu^{\lambda'}(k')\, \gamma_\mu \frac{1}{\hat{p}+\hat{k}'-im} e_\nu^\lambda(k)\ \gamma_\nu u^r(p) \right.$$
$$\left. + \bar{u}^{r'}(p')\, e_\mu^\lambda(k)\, \gamma_\mu \frac{1}{\hat{p}-\hat{k}'-im} e_\mu^{\lambda'}(k')\, \gamma_\nu u^r(p) \right]. \qquad (11.27a)$$

We now examine the operator $S^{(3)}$—the next term in the expansion of the S-matrix. This operator contains the T-product

$$T(A_\mu(x_1)\ A_\nu(x_2)\ A_\varrho(x_3)).$$

If this T-product is expanded in normal products, we obtain the term $N(A_\mu(x_1)\ A_\nu(x_2)\ A_\varrho(x_3))$ and the sum of all possible normal products with one contraction. Obviously the normal products with one contraction do not contribute to the matrix element of the process (as in the case examined earlier of one operator A). Let us look at the normal products of three electromagnetic field operators. If all three operators are creation operators, then acting to the left on the final state vector we obtain zero. If one of the operators is an annihilation operator, then the other two operators are creation operators and we again obtain zero. If two or three operators are annihilation operators, we obtain zero by acting on the initial state vector. Thus,

$$(\Phi^+_{p'r',\,k'\lambda'} S^{(3)} \Phi_{pr,\,k\lambda}) = 0. \qquad (11.28)$$

It is not difficult to show in an analogous manner that the operators $S^{(2m+1)}$ (m is any integer) do not contribute to the matrix element of the process (11.1).

Consider the operator $S^{(4)}$. A non-zero contribution to the matrix element of the process is made, in particular, by the following product of the electron–positron field operators:

$$\bar{\psi}^{(-)}(x_1)\ \gamma_\mu \overbrace{\psi(x_1)\ \bar{\psi}(x_2)}\ \gamma_\nu \overbrace{\psi(x_2)\ \bar{\psi}(x_3)}\ \gamma_\varrho \overbrace{\psi(x_3)\ \bar{\psi}(x_4)}\ \gamma_\sigma \psi^{(+)}(x_4). \qquad (11.29)$$

In the expansion of the original T-product there are 4! terms equivalent to (11.29).

We turn now to an examination of the normal products of electromagnetic field operators. It is obvious that a non-zero contribution to the matrix element of the process (11.1) can be made by the following normal products with one contraction:

$$\left.\begin{aligned} &N(\overbracket{A_\mu(x_1)\, A_\nu(x_2)\, A_\varrho(x_3)\, A_\sigma}(x_4)),\\ &N(\overbracket{A_\mu(x_1)\, A_\nu(x_2)\, A_\varrho}(x_3)\, A_\sigma(x_4)) \end{aligned}\right\} \tag{11.30}$$

and so forth. Furthermore, a non-zero contribution to the matrix element can be made only by the product of negative-frequency and positive-frequency components of operators, i.e.

$$\left.\begin{aligned} &\overbracket{A_\mu(x_1)\, A_\sigma}(x_4)\, [A_\nu^{(-)}(x_2)\, A_\varrho^{(+)}(x_3) + A_\varrho^{(-)}(x_3)\, A_\nu^{(+)}(x_2)],\\ &\overbracket{A_\mu(x_1)\, A_\varrho}(x_3)\, [A_\nu^{(-)}(x_2)\, A_\sigma^{(+)}(x_4) + A_\sigma^{(-)}(x_4)\, A_\nu^{(+)}(x_2)] \end{aligned}\right\} \tag{11.31}$$

and so forth.

To calculate the matrix element of the operator $S^{(4)}$ it is necessary to know the contraction of the electromagnetic field operators. Using commutation relations (7.40) and expansions (7.35) we obtain from the general expression (8.49)

$$\overbracket{A_\mu(x_1)\, A_\sigma}(x_4) = \begin{cases} \dfrac{1}{(2\pi)^3} \displaystyle\int \sum_{\lambda=1}^{4} e_\mu^\lambda(k)\, e_\sigma^\lambda(k)\, \eta_\lambda \frac{1}{2k_0}\, \mathrm{e}^{ik(x_1-x_4)}\, dk, & x_{10} > x_{40};\\ \dfrac{1}{(2\pi)^3} \displaystyle\int \sum_{\lambda=1}^{4} e_\sigma^\lambda(k)\, e_\mu^\lambda(k)\, \eta_\lambda \frac{1}{2k_0}\, \mathrm{e}^{-ik(x_1-x_4)}\, dk, & x_{40} > x_{10}. \end{cases} \tag{11.32}$$

The vectors $e^\lambda(k)$ form a complete set. From (7.15) it is obvious that

$$\sum_{\lambda=1}^{4} e_\mu^\lambda(k)\, e_\sigma^\lambda(k)\, \eta_\lambda = \delta_{\mu\sigma}. \tag{11.33}$$

Comparing (11.32) and (8.41) we conclude that

$$\overbracket{A_\mu(x_1)\, A_\sigma}(x_4) = \delta_{\mu\sigma}\, I(x_1 - x_4; 0) = \frac{-i}{(2\pi)^4}\, \delta_{\mu\sigma} \lim_{\varepsilon \to 0} \int \frac{\mathrm{e}^{ik(x_1-x_4)}}{k^2 - i\varepsilon}\, dk. \tag{11.34}$$

We now calculate the contribution to the Compton effect matrix element by the operator

$$\begin{aligned} (-i)^4 \int \bar\psi^{(-)}(x_1)\, (-ie)\, \gamma_\mu \overbracket{\psi(x_1)\, \bar\psi}(x_2)\, (-ie)\, \gamma_\nu \overbracket{\psi(x_2)\, \bar\psi}(x_3)\\ \times (-ie)\, \gamma_\varrho \overbracket{\psi(x_3)\, \bar\psi}(x)\, (-ie)\, \gamma_\sigma \psi^{(+)}(x_4)\, \overbracket{A_\mu(x_1)\, A_\sigma}(x)\\ \times A_\nu^{(-)}(x_2)\, A_\varrho^{(+)}(x_3)\, dx_1\, dx_2\, dx_3\, dx_4. \end{aligned} \tag{11.35}$$

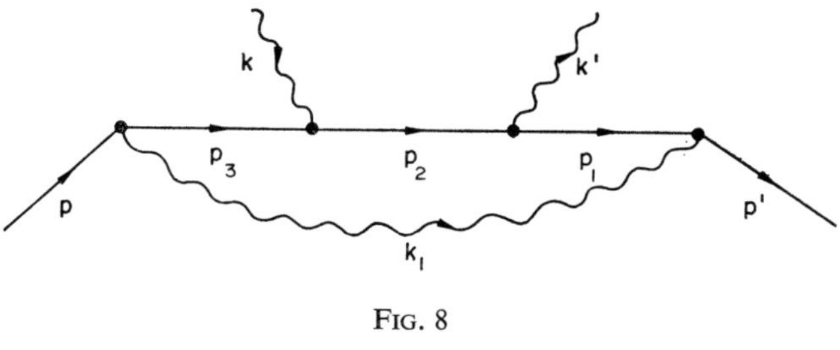

FIG. 8

Using (11.5), (11.12), (11.15), (11.34), and (8.58) and carrying out the integration over the variables $x_1, x_2, \ldots$, we obtain

$$
\begin{aligned}
(\Phi^{+}_{p'r',\,k'\lambda'} S^{(4)} \Phi_{pr,\,k\lambda}) = (-i)^4 \int \Bigg[& \frac{1}{(2\pi)^{3/2}} \left(\frac{m}{p_0'}\right)^{1/2} \bar{u}^{r'}(p')(-ie)\gamma_\mu (2\pi)^4 \delta(p'-p_1-k_1) \\
& \times \frac{(-1)}{(2\pi)^4} \frac{1}{\hat{p}_1 - im} (-ie)\gamma_\nu (2\pi)^4 \delta(p_1+k'-p_2) \frac{(-1)}{(2\pi)^4} \\
& \times \frac{1}{\hat{p}_2 - im} (-ie)\gamma_\varrho (2\pi)^4 \delta(p_2-k-p_3) \\
& \times \frac{(-1)}{(2\pi)^4} \frac{1}{\hat{p}_3 - im} (-ie)\gamma_\sigma (2\pi)^4 \delta(p_3-p+k_1) u^r(p) \left(\frac{m}{p_0}\right)^{1/2} \\
& \times \frac{1}{(2\pi)^{3/2}} \frac{(-i)}{(2\pi)^4} \delta_{\mu\sigma} \frac{1}{k_1^2} \frac{1}{(2\pi)^{3/2}} \frac{1}{\sqrt{2k_0'}} e_\nu^{\lambda'}(k') \\
& \times \frac{1}{(2\pi)^{3/2}} \frac{1}{\sqrt{2k_0}} e_\varrho^{\lambda}(k)\, dp_1\, dp_2\, dp_3\, dk_1 \Bigg] + \ldots . \qquad (11.36)
\end{aligned}
$$

The calculated term of the matrix element of the operator $S^{(4)}$ corresponds to the diagram in Fig. 8. In comparison to the diagrams examined earlier this diagram has a new element —an internal wavy line. It is obvious that this new element in the diagram is related to the contraction of the electromagnetic field operators. We put

$$\frac{(-i)}{(2\pi)^4} \frac{1}{k_1^2} \delta_{\mu\sigma}$$

in correspondence with the internal wavy line of the diagram, where k_1 is the four-momentum of the internal wavy line; μ and σ are indices of the matrices at the vertices united by this line. For all the remaining elements we take the same rules of correspondence as for the diagrams in Figs. 6 and 7. If, moving opposite the arrows along the solid line in Fig. 8, we write out the quantities to which the lines and vertices correspond, then multiply the resulting expression by the quantities to which the wavy lines correspond, and integrate that expression over the momenta of all internal lines, we obtain the calculated term of the matrix element.

The diagram in Fig. 8 describes the following sequence of absorption and emission events. The initial electron emits a virtual photon with momentum k_1 and goes into the intermediate state with momentum p_3. Then the virtual electron absorbs the initial photon and goes into the intermediate state with momentum p_2. The virtual electron with momentum p_2 emits the final photon and goes into the intermediate state with momentum p_1. Finally, the virtual electron with momentum p_1 absorbs the virtual photon and goes into the final state. The quantity

$$\frac{(-i)}{(2\pi)^4} \frac{1}{k_1^2} \delta_{\mu\sigma}$$

describes the propagation of the virtual photon and is called the *photon propagator*.

It is obvious that the second term in the square brackets in (11.31) reduces to the component of the matrix element to which the diagram in Fig. 9 corresponds (first the virtual photon is emitted, then the final photon is emitted, the initial photon is absorbed, and finally the virtual photon is absorbed).

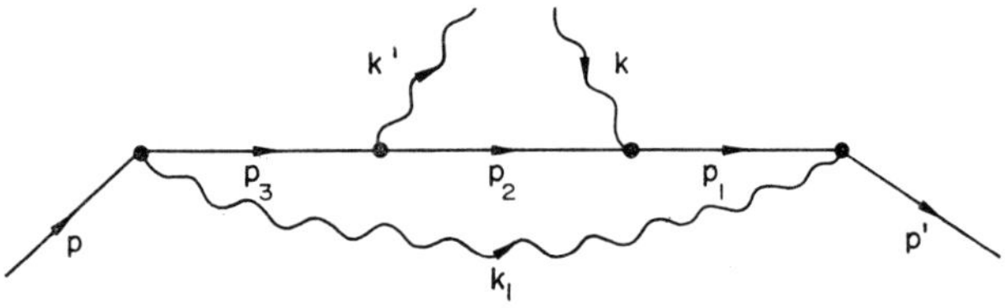

FIG. 9

It is clear from the above that the operator (11.29) corresponds to the diagram skeleton in Fig. 10. The various terms in the expansion of the T-product of electromagnetic field operators correspond to the different possible manners of attaching the initial, final, and virtual photons to the vertices in Fig. 10. Note that the formal integration of the expressions corresponding to these diagrams yields infinity. In order to calculate the true contribution of fourth-order terms, it is necessary to renormalize the mass and charge of the electron. After renormalization the contribution of these terms turns out to be finite. We shall not discuss here the questions related to renormalization. We examined the matrix element of the operator $S^{(4)}$ only in order to show how to construct diagrams of an order higher than the first non-vanishing one.

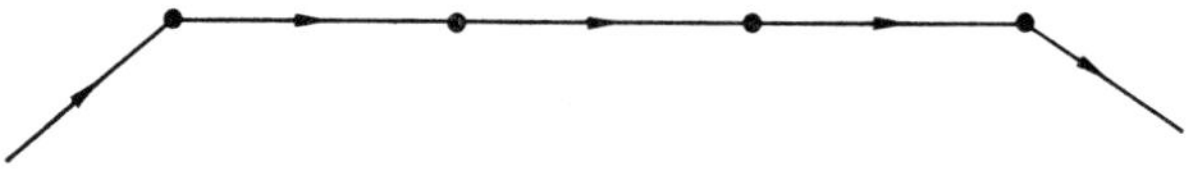

FIG. 10

ELECTRON–ELECTRON SCATTERING

For our next example we shall look at the process of electron scattering by a free electron

$$e+e \rightarrow e+e. \tag{11.37}$$

Let p_1, p_2, and p_1', p_2' designate the four-momenta of the initial and final electrons respectively. The vectors describing the initial and final states are

$$\begin{aligned} \Phi_{p_1, p_2} &= c^+(p_1)\, c^+(p_2)\Phi_0, \\ \Phi_{p_1', p_2'} &= c^+(p_1')\, c^+(p_2')\,\Phi_0. \end{aligned} \tag{11.38}$$

The spin indices $r_1, r_2, \ldots$ have been omitted here. Henceforth, where spin indices are not explicitly written out, they will be implied.

Our task is to calculate the matrix element

$$(\Phi^+_{p_1', p_2} S\Phi_{p_1, p_2}) \tag{11.39}$$

in the lowest order of perturbation theory. It is obvious that the operator $S^{(1)}$ does not contribute to the matrix element of the process (11.37).

Let us examine the operator $S^{(2)}$. In the expansion of the T-product in normal products one needs to take into account only the term

$$N(\bar{\psi}^{(-)}(x_1)\,\gamma_\mu\psi^{(+)}(x_1)\,\bar{\psi}^{(-)}(x_2)\,\gamma_\nu\psi^{(+)}(x_2))\,\overbrace{A_\mu(x_1)\,A_\nu(x_2)}. \tag{11.40}$$

Actually, the matrix element can be different from zero only if there are two positive-frequency operators in the product of fermion operators (if the number of positive-frequency operators is greater than two, acting to the left on the initial vector results in zero; if there are less than two, acting to the left on the final state vector results in zero). The product of two positive-frequency operators occurs in the combinations $\psi^{(+)}\psi^{(+)}$, $\bar{\psi}^{(+)}\psi^{(+)}$, $\bar{\psi}^{(+)}\bar{\psi}^{(+)}$. Since there are no positrons in the initial state, it is obvious that only the product containing the first of the terms written out makes a non-zero contribution. Considering also that there are no photons in the initial and final states, we conclude that the operator (11.40) makes a non-zero contribution to the matrix element.

We act on the initial state vector with the operator $\psi^{(+)}_{\beta_1}(x_1)\,\psi^{(+)}_{\beta_2}(x_2)$. Using (10.8) and (10.10) we have

$$\begin{aligned}\psi^{(+)}_{\beta_1}(x_1)\,\psi^{(+)}_{\beta_2}(x_2)\,c^+(p_1)\,c^+(p_2)\Phi_0 &= \psi^{(+)}_{\beta_1}(x_1)\,([\psi^{(+)}_{\beta_2}(x_2),\,c^+(p_1)]_+ \\ -c^+(p_1)\,\psi^{(+)}_{\beta_2}(x_2))\,c^+(p_2)\Phi_0 &= \frac{1}{(2\pi)^3}\left(\frac{m^2}{p_{10}p_{20}}\right)^{1/2} \\ \times[u_{\beta_2}(p_1)\,e^{ip_1x_2}u_{\beta_1}(p_2)\,e^{ip_2x_1}-u_{\beta_2}(p_2)\,e^{ip_2x_2}u_{\beta_1}(p_1)\,e^{ip_1x_1}]\,\Phi_0. \end{aligned} \tag{11.41}$$

Thus, if we act on the vector describing two electrons with momenta p_1 and p_2 with the operator $\psi^{(+)}(x_1)\,\psi^{(+)}(x_2)$, we obtain the vector Φ_0 which is multiplied by the antisymmetrized product (due to the indistinguishability of the electrons) of the corresponding solutions of the free Dirac equation.

We now act with the operator $\bar{\psi}^{(-)}_{\alpha_1}(x_1)\,\bar{\psi}^{(-)}_{\alpha_2}(x_2)$ to the left on the final state vector. Using (10.13) and (10.15) we find

$$\begin{aligned}\Phi_0^+\,c(p_2')\,c(p_1')\,\bar{\psi}^{(-)}_{\alpha_1}(x_1)\bar{\psi}^{(-)}_{\alpha_2}(x_2) &= \Phi_0^+\,c(p_2')\,([c(p_1'),\,\bar{\psi}^{(-)}_{\alpha_1}(x_1)]_+-\bar{\psi}^{(-)}_{\alpha_1}(x_1)\,c(p_1'))\,\bar{\psi}^{(-)}_{\alpha_2}(x_2) \\ =\Phi_0^+\frac{1}{(2\pi)^3}\left(\frac{m^2}{p_{10}'p_{20}'}\right)^{1/2}&[\bar{u}_{\alpha_2}(p_2')\,e^{-ip_2'x_2}\,\bar{u}_{\alpha_1}(p_1')\,e^{-ip_1'x_1}-\bar{u}_{\alpha_2}(p_1')\,e^{-ip_1'x_2}\bar{u}_{\alpha_1}(p_2')\,e^{-ip_2'x_1}]. \end{aligned} \tag{11.42}$$

Obviously the operator (11.40) is equal to

$$-\bar{\psi}^{(-)}_{\alpha_1}(x_1)\,\bar{\psi}^{(-)}_{\alpha_2}(x_2)\,\psi^{(+)}_{\beta_1}(x_1)\,\psi^{(+)}_{\beta_2}(x_2)(\gamma_\mu)_{\alpha_1\beta_1}\,(\gamma_\nu)_{\alpha_2\beta_2}\overbrace{A_\mu(x_1)A_\nu(x_2)}. \tag{11.43}$$

Let us calculate the matrix element of this operator. Using (11.41), (11.42), and (11.34) and integrating over the variables x_1 and x_2 we obtain†

$$\begin{aligned}(\Phi^+_{p_1',p_2'}S^{(2)}\Phi_{p_1,p_2}) &= \frac{(-i)^2}{2!}\,2\frac{(-i)}{(2\pi)^4}\,\delta_{\mu\nu}\,\frac{1}{(2\pi)^6}\left(\frac{m^4}{p_{10}p_{20}p_{10}'p_{20}'}\right)^{1/2} \\ \times(-ie)^2(2\pi)^8\int&[(\bar{u}(p_1')\,\gamma_\mu\delta(p_1'-p_1-k)\,u(p_1))(\bar{u}(p_2')\,\gamma_\nu\delta(p_2'-p_2+k)\,u(p_2)) \\ -(\bar{u}(p_2')\;\gamma_\mu\delta(p_2'-p_1-k)\;u(p_1))&(\bar{u}(p_1')\;\gamma_\nu\delta(p_1'-p_2+k)\;u(p_2))]\frac{1}{k^2}\,dk. \end{aligned} \tag{11.44}$$

† In multiplying (11.42) and (11.41) four terms arise. It is not difficult to verify that they are pairwise equal. The factor of 2 in the expression (11.44) is related to this.

Let us construct the Feynman diagram corresponding to the matrix element (11.44). We put in correspondence with the first term in (11.44) the diagram in Fig. 11. The correspondence rules for the lines and vertices of this diagram and the parts of the matrix element are the same as we used before. In order to obtain the first term of the matrix element of the diagram in Fig. 11, it is necessary, moving opposite the arrows along one of the solid lines, to write out all expressions which correspond to the encountered lines and vertices, then to multiply the resulting expression by the expression which arises as a result of the movement opposite the arrows along the second solid line, and finally to multiply the whole expression by the photon propagator and integrate over the momentum of the virtual photon.The diagram describes the following interaction events. The initial electron with momentum p_2 emits a virtual photon with momentum k and goes into the final state with momentum p_2'. The other initial electron whose momentum is p_1 absorbs this virtual photon and goes into the final state with momentum† p_1'.

The second term in the matrix element (11.44) arises as a result of the indistinguishability of the electrons. The diagram in Fig. 12 obviously corresponds to it. Thus, because of Pauli's principle, besides the first possibility described by the diagram in Fig. 11, it is also necessary to take into account the possibility of the transition of electrons from the state with momenta p_2 and p_1 into the states with momenta p_1' and p_2' respectively (see Fig. 12) and to take the difference of the matrix elements to which the diagrams in Figs. 11 and 12 correspond.

From the anticommutation relations of the electron creation operators it follows that the matrix element (11.39) must be antisymmetric under the transposition of p_1 and p_2 or p_1' and p_2' (we actually are considering the transposition of p_1r_1 and p_2r_2 or $p_1'r_1'$ and $p_2'r_2'$). The

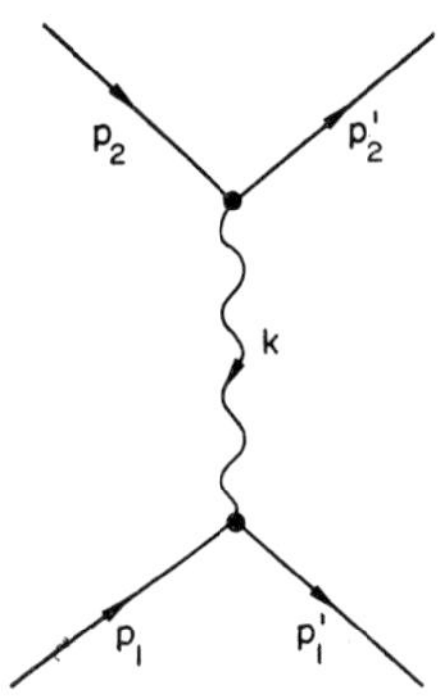

FIG. 11

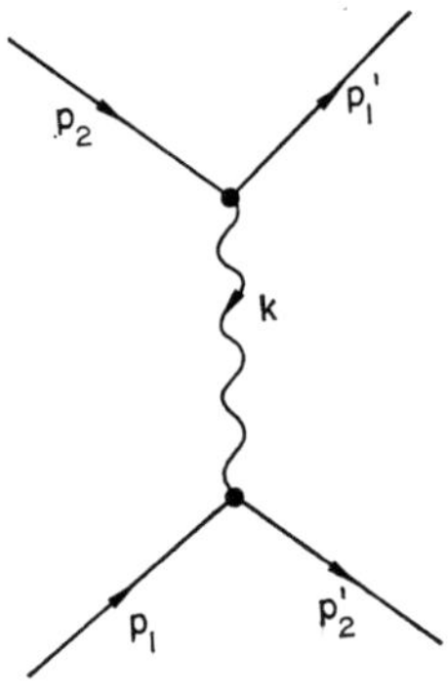

FIG. 12

difference of the matrix elements given by the diagrams in Figs. 11 and 12 has, as is easy to see, the required properties of antisymmetry.

In the expression (11.44) one can integrate over k and sum over ν. Ultimately we find to

† If the direction of the arrow on the internal photon line in the diagram in Fig. 11 is reversed, this will correspond to a reversal $k \to -k$ in the first term of the integral (11.44). Thus, the matrix element does not change if the direction of the arrow on the internal photon line is reversed.

second order in e the matrix element of the electron–electron scattering process is equal to

$$(\Phi^{+}_{p_1', p_2'} S^{(2)} \Phi_{p_1 p_2}) = (-i)^2 \frac{(-i)}{(2\pi)^4} \frac{1}{(2\pi)^6} \left(\frac{m^4}{p_{10} p_{20} p_{10}' p_{20}'} \right)^{1/2} (-ie)^2 (2\pi)^8$$

$$\times \left[\frac{(\bar{u}(p_1')\,\gamma_\mu u(p_1))(\bar{u}(p_2')\,\gamma_\mu u(p_2))}{(p_1'-p_1)^2} - \frac{(\bar{u}(p_2')\,\gamma_\mu u(p_1))(\bar{u}(p_1')\,\gamma_\mu u(p_2))}{(p_2'-p_1)^2} \right] \delta(p_1'+p_2'-p_1-p_2). \tag{11.45}$$

It is clear that $\delta(p_1'+p_2'-p_1-p_2)$ ensures the conservation of the total energy and the total momentum.

PROCESSES IN WHICH POSITRONS TAKE PART

In this subsection we shall examine processes in which, in addition to electrons and photons, positrons take part.

Let us consider the operator

$$N(\bar{\psi}(x_1)\,\gamma_\mu \overbrace{\psi(x_1)\,\bar{\psi}(x_2)}\,\gamma_\nu \psi(x_2))\, N(A_\mu(x_1)\, A_\nu(x_2)) \tag{11.46}$$

and ascertain to which processes it contributes. We separate the spinor operators into positive- and negative-frequency parts:

$$N(\bar{\psi}(x_1)\,\gamma_\mu \overbrace{\psi(x_1)\,\bar{\psi}(x_2)}\,\gamma_\nu \psi(x_2)) = \bar{\psi}^{(+)}(x_1)\,\gamma_\mu \overbrace{\psi(x_1)\,\bar{\psi}(x_2)}\,\gamma_\nu \psi^{(+)}(x_2)$$

$$+\bar{\psi}^{(-)}(x_1)\,\gamma_\mu \overbrace{\psi(x_1)\,\bar{\psi}(x_2)}\,\gamma_\nu \psi^{(-)}(x_2) + \bar{\psi}^{(-)}(x_1)\,\gamma_\mu \overbrace{\psi(x_1)\,\bar{\psi}(x_2)}\,\gamma_\nu \psi^{(+)}(x_2)$$

$$-\psi^{(-)}(x_2)(\gamma_\mu \overbrace{\psi(x_1)\,\bar{\psi}(x_2)}\,\gamma_\nu)\,\bar{\psi}^{(+)}(x_1). \tag{11.47}$$

As we have seen, the third term on the right-hand side of (11.47) contributes to the Compton effect. Let us look at the operator

$$\bar{\psi}^{(+)}(x_1)\,\gamma_\mu \overbrace{\psi(x_1)\,\bar{\psi}(x_2)}\,\gamma_\nu \psi^{(+)}(x_2)\, N(A_\mu(x_1)\, A_\nu(x_2)). \tag{11.48}$$

The matrix element of this operator is zero unless there is an electron and a positron in the initial state. Two electromagnetic field operators can annihilate two initial or two final photons or one initial and one final photon. The various processes relate to these possibilities. Taking into account, however, the law of conservation of four-momentum, we conclude that the two photons must be in the final state, i.e. the operator (11.48) contributes to the matrix element of the process of the conversion of the electron–positron pair into two photons:

$$\bar{e}+e \rightarrow \gamma+\gamma. \tag{11.49}$$

It is obvious that none of the remaining operators in $S^{(2)}$ contribute to the matrix element of this process. We denote the four-momenta of the initial electron and positron respectively by p and p' and the four-momenta of the final photons by k and k'. The initial and final state vectors are respectively

$$\Phi_{p,\,p'} = c^+(p)\, d^+(p')\, \Phi_0, \quad \Phi_{k,\,k'} = a^+(k)\, a^+(k')\, \Phi_0. \tag{11.50}$$

We act with the operator $\bar{\psi}_\alpha^{(+)}(x_1)\,\psi_\beta^{(+)}(x_2)$ on the initial vector $\Phi_{p,\,p'}$. Using (10.10), expansion (8.16), commutation relations (6.40), and conditions (6.46) we obtain

$$\bar{\psi}_\alpha^{(+)}(x_1)\,\psi_\beta^{(+)}(x_2)\,c^+(p)\,d^+(p')\,\Phi_0 = \frac{1}{(2\pi)^{3/2}}\left(\frac{m}{p_0}\right)^{1/2} u_\beta(p)\,\mathrm{e}^{ipx_2}\,([\bar{\psi}_\alpha^{(+)}(x_1),\,d^+(p')]_+$$

$$-\,d^+(p')\,\bar{\psi}_\alpha^{(+)}(x_1))\,\Phi_0 = \frac{1}{(2\pi)^3}\left(\frac{m^2}{p_0p_0'}\right)^{1/2} u_\beta(p)\,\mathrm{e}^{ipx_2}\,\bar{u}_\alpha(-p')\,\mathrm{e}^{ip'x_1}\,\Phi_0. \quad (11.51)$$

In the expansion of the operator $N(A_\mu(x_1)A_\nu(x_2))$ a non-zero contribution to the matrix element of the process (11.49) is made by the term $A_\mu^{(-)}(x_1)A_\nu^{(-)}(x_2)$. We act with this operator to the left on the final vector $\Phi_{k,\,k'}^+$. Using (11.15) we find (see the remark on p. 86)

$$\Phi_0^+a(k)\,a(k')\,A_\mu^{(-)}(x_1)\,A_\nu^{(-)}(x_2) = \Phi_0^+a(k)([a(k'),A_\mu^{(-)}(x_1)]+A_\mu^{(-)}(x_1)\,a(k'))\,A_\nu^{(-)}(x_2)$$

$$= \Phi_0^+\frac{1}{(2\pi)^3}\frac{1}{\sqrt{4k_0k_0'}}\,[e_\mu(k')\,\mathrm{e}^{-ik'x_1}\,e_\nu(k)\,\mathrm{e}^{-ikx_2}+e_\mu(k)\,\mathrm{e}^{-ikx_1}\,e_\nu(k')\,\mathrm{e}^{-ik'x_2}]. \quad (11.52)$$

Two terms arise as a result of the indistinguishability of the photons.

With the help of (11.51) and (11.52) we obtain the following expression for the matrix element of the process (11.49) in the second order of perturbation theory in e:

$$(\Phi_{k,\,k'}^+S^{(2)}\Phi_{p,\,p'}) = (-i)^2\frac{1}{(2\pi)^6}\left(\frac{m^2}{p_0p_0'}\right)^{1/2}\frac{1}{\sqrt{4k_0k_0'}}\frac{(-1)}{(2\pi)^4}(2\pi)(-ie)^2$$

$$\times\int\Big[\bar{u}(-p')\,\gamma_\mu\delta(-p'+k'-p_1)\,\frac{1}{\hat{p}_1-im}\,\gamma_\nu\delta(p_1-p-(-k))\,u(p)\,e_\mu(k')\,e_\nu(k)$$

$$+\bar{u}(-p')\,\gamma_\mu\delta(-p'-p_1-(-k))\,\frac{1}{\hat{p}_1-im}\,\gamma_\nu\delta(p_1+k'-p)\,u(p)\,e_\mu(k)\,e_\nu(k')\Big]\,dp_1. \quad (11.53)$$

This matrix element has the same structure as the matrix element of the Compton effect [expression (11.27)]. We put into correspondence with the first term in (11.53) the diagram in Fig. 13 (cf. the diagram in Fig. 6). If we assume the same rules of correspondence as before for the vertices, internal lines, entering solid lines with momentum p, and emerging wavy lines with momentum k', and put into correspondence with the emerging line with momentum $-p'$ the spinor

$$\frac{1}{(2\pi)^{3/2}}\left(\frac{m}{p_0'}\right)^{1/2}\bar{u}(-p'),$$

and with the entering wavy line with momentum $-k$ the vector

$$\frac{1}{(2\pi)^{3/2}}\frac{1}{\sqrt{2k_0}}\,e_\nu(k),$$

then it is obvious that the first term of the matrix element (11.53) can be obtained from the diagram in Fig. 13 in the usual manner. The second term of the matrix element (11.53) corresponds to the diagram in Fig. 14.

Comparing the diagrams in Figs. 6 and 7 with the diagrams in Figs. 13 and 14 we come to the following conclusion. If the emerging electron line in Figs. 6 and 7 is given the momentum $-p'$ $(-p_0' < 0)$ and the entering photon line is given $-k$ $(-k_0 < 0)$, then the corresponding diagrams describe the process of converting an electron with momentum p and a positron with momentum p' into two photons with momenta k and k'. Thus, the emission of an electron with the unphysical four-momentum $-p'$ $(-p_0' < 0)$ describes the absorption of a positron with physical four-momentum p' and the absorption of a photon with unphysical four-momentum $-k$ $(-k_0 < 0)$ describes the emission of a photon with momentum k. The direction in which the process develops is indicated in Figs. 13 and 14 by double arrows (the electron with momentum p emits a photon and goes into the intermediate state with momentum p_1; the positron with momentum p' annihilates with the virtual electron and forms the second final photon).

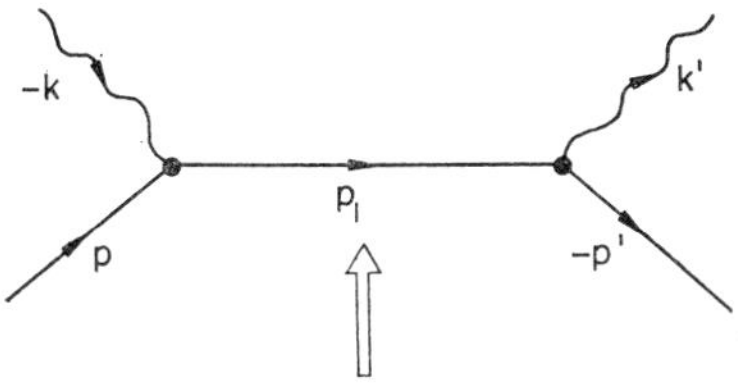

FIG. 13

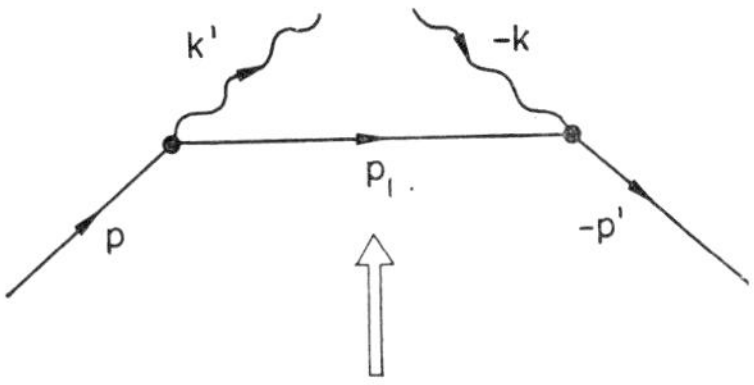

FIG. 14

We integrate over p_1 in (11.53). Ultimately the matrix element of the process of annihilation of the electron–positron pair into two photons is

$$(\Phi^+_{k,k'} S^{(2)} \Phi_{p,p'}) = \frac{-e^2}{(2\pi)^2} \left(\frac{m^2}{4p_0p_0'k_0k_0'} \right)^{1/2} \left[\bar{u}(-p')\,\hat{e}(k') \frac{1}{\hat{p}-\hat{k}-im}\,\hat{e}(k)\,u(p) \right.$$
$$\left. + \bar{u}(-p')\hat{e}(k) \frac{1}{\hat{p}-\hat{k}'-im}\,\hat{e}(k')\,u(p) \right] \delta(k+k'-p-p'), \qquad (11.54)$$

where $\hat{e}(k) = e_\mu(k)\,\gamma_\mu$.

We turn now to an examination of the following term in the expansion (11.47):

$$\bar{\psi}^{(-)}(x_1)\,\gamma_\mu \overline{\psi(x_1)\,\bar{\psi}}(x_2)\,\gamma_\nu \psi^{(-)}(x_2)\,N(A_\mu(x_1)\,A_\nu(x_2)). \qquad (11.55)$$

The matrix element of this operator is zero unless there is an electron and a positron in the final state. From the law of conservation of the total four-momentum it furthermore follows that there must be two photons in the initial state. Thus, the operator (11.55) contributes to the matrix element of the process:

$$\gamma+\gamma \rightarrow e+\bar{e}. \qquad (11.56)$$

Let us denote the four-momenta of the final positron and electron by p and p' and the four-momenta of the initial photons by k and k'. The initial and final state vectors are

$$\left.\begin{aligned} \Phi_{k,k'} &= a^+(k)\,a^+(k')\,\Phi_0, \\ \Phi_{p,p'} &= c^+(p')\,d^+(p)\,\Phi_0. \end{aligned}\right\} \qquad (11.57)$$

We have

$$\left.\begin{aligned}
&\Phi_0^+ d(p)\, c(p')\, \bar{\psi}_\alpha^{(-)}(x_1)\, \psi_\beta^{(-)}(x_2)\\
&\quad = \Phi_0^+ \frac{1}{(2\pi)^3}\left(\frac{m^2}{p_0 p_0'}\right)^{1/2} \bar{u}_\alpha(p')\, e^{-ip'x_1} u_\beta(-p)\, e^{-ipx_2},\\
&A_\mu^{(+)}(x_1)\, A_\nu^{(+)}(x_2)\, a^+(k)\, a^+(k')\, \Phi_0\\
&\quad = \frac{1}{(2\pi)^3}\frac{1}{\sqrt{4k_0k_0'}}[e_\nu(k)\, e^{ikx_2}\, e_\mu(k')\, e^{ik'x_1}\\
&\qquad + e_\nu(k')\, e^{ik'x_2} e_\mu(k)\, e^{ikx_1}]\, \Phi_0.
\end{aligned}\right\} \tag{11.58}$$

With the help of (11.58) we find for the matrix element of the process (11.56) in the second order of perturbation theory the expression

$$\begin{aligned}
(\Phi_{p,p'}^+ S^{(2)} \Phi_{k,k'}) &= (-i)^2 \frac{1}{(2\pi)^6}\left(\frac{m^2}{4p_0p_0'k_0k_0'}\right)^{1/2} \frac{(-1)}{(2\pi)^4}(-ie)^2 (2\pi)^8\\
&\times \int [\bar{u}(p')\, \gamma_\mu \delta(p'+(-k')-p_1)\frac{1}{\hat{p}_1 - im}\gamma_\nu\, \delta(p_1-k-(-p))\, u(-p)\\
&\times e_\mu(k')\, e_\nu(k) + \bar{u}(p')\, \gamma_\mu\, \delta(p'-p_1-k)\frac{1}{\hat{p}_1-im}\\
&\times \gamma_\nu\, \delta(p_1+(-k')-(-p))\, u(-p)\, e_\mu(k)\, e_\nu(k')]\, dp_1.
\end{aligned} \tag{11.59}$$

If we put the entering solid line with four-momentum $-p$ into correspondence with the spinor

$$\frac{1}{2\pi}\left(\frac{m}{p_0}\right)^{1/2} u(-p),$$

the emerging wavy line with momentum $-k'$ with the vector

$$\frac{1}{(2\pi)^{3/2}}\frac{1}{\sqrt{2k_0'}}e(k'),$$

and assume all previous rules of correspondence for the remaining quantities in the matrix element (11.59), then it is obvious that this matrix element corresponds to the Feynman diagram depicted in Fig. 15.

Comparing the diagrams in Fig. 15 and the diagrams in Figs. 6 and 7 we come to the following conclusion. If in the diagrams in Figs. 6 and 7 the four-momentum of the entering

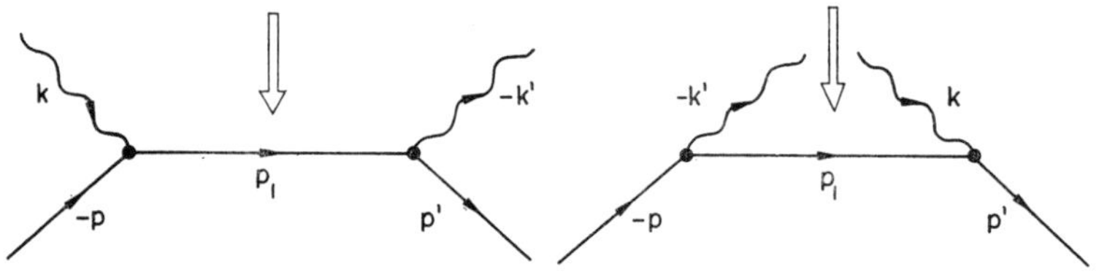

FIG. 15

electron is set equal to $-p$ $(-p_0 < 0)$ and the momentum of the emerging photon line is set equal to $-k'$ $(-k_0' < 0)$, then these diagrams describe the process of the conversion of two photons with momenta k and k' into an electron with momentum p' and a positron with momentum p. Thus, the entering electron line with the unphysical four-momentum $-p$ $(-p_0 < 0)$ corresponds to the emitted positron with the physical four-momentum p and the emerging photon line with the unphysical four-momentum $-k'$ $(-k_0' < 0)$ corresponds to the absorption of the photon with four-momentum k'. The direction in which the physical process develops is shown by the double arrows in Fig. 16 (the photon forms a positron with momentum p and a virtual electron with momentum p_1; the virtual electron absorbs the second photon and goes into the final state with momentum p').

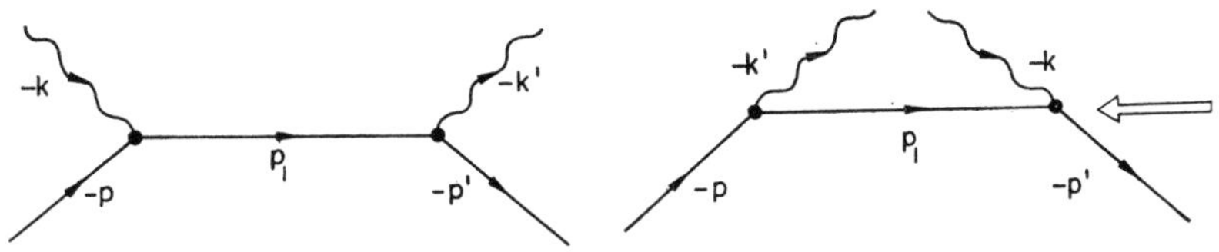

FIG. 16

Integrating (11.59) over p_1 we ultimately obtain for the matrix element of the process (11.56)

$$(\Phi^+_{p,\,p'} S^{(2)} \Phi_{k,\,k'}) = \frac{-e^2}{(2\pi)^2} \left(\frac{m^2}{4 p_0 p_0' k_0 k_0'} \right)^{1/2} \delta(p'+p-k-k')$$

$$\times \left[\bar{u}(p')\hat{e}(k') \frac{1}{-\hat{p}+\hat{k}-im} \hat{e}(k)\, u(-p) \right.$$

$$\left. + \bar{u}(p')\, \hat{e}(k) \frac{1}{-\hat{p}+\hat{k}'-im} \hat{e}(k')\; \bar{u}(-p) \right]. \tag{11.60}$$

Finally, let us examine the operator

$$-\psi^{(-)}(x_2)\,(\gamma_\mu \psi(x_1)\, \bar{\psi}(x_2)\, \gamma_\nu)\, \bar{\psi}^{(+)}(x_1)\, N(A_\mu(x_1)\, A_\nu(x_2)). \tag{11.61}$$

The matrix element of this operator is zero unless there is a positron and a photon in the initial and final states. Thus, the operator (11.61) contributes to the matrix element of the process of photon–positron scattering:

$$\gamma + \bar{e} \rightarrow \gamma + \bar{e}. \tag{11.62}$$

Denote the four-momenta of the initial and final photons (positrons) respectively by k' and k (p' and p). The initial and final state vectors are

$$\left.\begin{aligned} \Phi_{p',\,k'} &= d^+(p')\, a^+(k')\Phi_0, \\ \Phi_{p,\,k} &= d^{(+)}(p)\, a^+(k)\, \Phi_0. \end{aligned}\right\} \tag{11.63}$$

We have

$$\left.\begin{aligned}\bar{\psi}_\alpha^{(+)}(x_1)\, d^+(p')\, a^+(k')\Phi_0 &= \frac{1}{(2\pi)^{3/2}}\left(\frac{m}{p_0'}\right)^{1/2} \bar{u}_\alpha(-p')\, e^{ip'x_1} a^+(k')\,\Phi_0, \\ \Phi_0^+ a(k)\, d(p)\, \psi_\beta^{(-)}(x_2) &= \Phi_0^+ a(k) \frac{1}{(2\pi)^{3/2}}\left(\frac{m}{p_0}\right)^{1/2} u_\beta(-p)\, e^{-ipx_2}.\end{aligned}\right\} \qquad (11.64)$$

Using (11.15), (11.12), and (11.64) we obtain for the matrix element of the process of photon–positron scattering in the second order of perturbation theory

$$\begin{aligned}(\Phi_{p,k}^+ S^{(2)}\Phi_{p',k'} &= (-1)(-i)^2 \frac{1}{(2\pi)^6}\left(\frac{m^2}{4p_0p_0'k_0k_0'}\right)^{1/2} \\ &\times \frac{(-1)}{(2\pi)^4}(2\pi)^8(-ie)^2 \int [\bar{u}(-p')\,\gamma_\mu\,\delta((-p')+(-k')-p_1)\frac{1}{\hat{p}_1-im}\gamma_\nu \\ &\times \delta(p_1-(-p)-(-k))\, u(-p)\, e_\mu(k')\, e_\nu(k) + \bar{u}(-p')\,\gamma_\mu\,\delta((-p')-p_1-(-k)) \\ &\times \frac{1}{\hat{p}_1-im}\gamma_\nu\,\delta(p_1+(-k')-(-p))\, u(-p)\, e_\mu(k)\, e_\nu(k')]\, dp_1. \end{aligned} \qquad (11.65)$$

Obviously the matrix element (11.65) corresponds to the Feynman diagram in Fig. 16. Comparing the diagrams in Figs. 6, 7, and 16 we come to the following conclusion. If in the diagrams in Figs. 6 and 7 one sets the momenta of the entering and emerging electron lines equal respectively to $-p$ $(-p_0 < 0)$ and $-p'$ $(-p_0' < 0)$ and the momenta of the entering and emerging photon lines equal respectively to $-k$ and $-k'$ $(-k_0 < 0,\ -k_0' < 0)$, then the diagrams correspond to the process of photon–positron scattering (the momenta of the initial and final positrons are p' and p and the momenta of the initial and final photons are k' and k).

Integrating (11.65) over the momentum p_1 we ultimately obtain for the matrix element of the process of photon–positron scattering

$$\begin{aligned}(\Phi_{p,k}^+ S^{(2)}\,\Phi_{p',k'}) &= \frac{1}{(2\pi)^2}e^2\left(\frac{m^2}{4p_0p_0'k_0k_0'}\right)^{1/2}\delta(p+k-p'-k')\times \\ &\times\left[\bar{u}(-p')\,\hat{e}(k')\frac{1}{-\hat{p}-\hat{k}-im}\hat{e}(k)\, u(-p)\right. \\ &\left.+\bar{u}(-p')\,\hat{e}(k')\frac{1}{-\hat{p}+\hat{k}'-im}\hat{e}(k')\, u(-p)\right].\end{aligned} \qquad (11.66)$$

Thus, if one recognizes that the momenta of the external lines in Feynman diagrams can take unphysical values (with a negative zero component) as well as physical, then the same diagrams (Figs. 6 and 7) describe the following four processes:

$$\left.\begin{aligned}\gamma+e &\to \gamma+e, \\ e+\bar{e} &\to \gamma+\gamma, \\ \gamma+\gamma &\to e+\bar{e}, \\ \gamma+\bar{e} &\to \gamma+\bar{e}.\end{aligned}\right\} \qquad (11.67)$$

A general rule which can be formulated on the basis of the discussion presented here consists in the following: the emerging (entering) line of a particle with an unphysical momentum $-p$ $(-p_0 < 0)$ corresponds to the absorption (emission) of an antiparticle with four-momentum p (the photon is a truly neutral particle; the emerging photon line with four-momentum $-k$ $(-k_0 < 0)$ describes the absorption of a photon with momentum k). This rule is valid for diagrams of any order and for any kind of interaction.

12. Currents. Form Factors

In this section we shall examine processes in which leptons (electrons, μ-mesons, neutrinos) and hadrons (nucleons, π-mesons, and others) take part. Leptons participate only in electromagnetic and weak interactions; hadrons participate in weak, electromagnetic, and strong interactions. Weak and electromagnetic interactions can be examined through perturbation theory. Perturbation theory is not applicable to strong interactions (the coupling constant is too large). We shall consider the strong interactions of hadrons phenomenologically by introducing form factors.

We begin with the process of electron scattering by nucleons:

$$e+N \rightarrow e+N. \tag{12.1}$$

The full interaction Hamiltonian is written in the form

$$\mathcal{H}_I(x) = \mathcal{H}_I^e(x)+\mathcal{H}_I^{ea}(x)+\mathcal{H}_I^a(x). \tag{12.2}$$

Here

$$\mathcal{H}_I^e(x) = -ieN(\bar{\psi}_e(x)\, \gamma_\mu \psi_e(x))\, A_\mu(x)$$

is the Hamiltonian of the interaction between electromagnetic and electron–positron fields; $H_I^{ea}(x)$ is the Hamiltonian of the interaction between the electromagnetic field and fields of all strongly interacting particles, and $H_I^a(x)$ is the Hamiltonian of all strong interactions.

The Hamiltonian $H_I^{ea}(x)$ has the form

$$\mathcal{H}_I^{ea}(x) = ej_\mu^p(x)\, A_\mu(x) - eN(j_\mu^\pi(x)\, A_\mu(x)) - e^2N(\phi^+(x)\, \phi(x)\, A_\mu(x)\, A_\mu(x)) + \ldots, \tag{12.3}$$

where

$$\left.\begin{aligned} j_\mu^p(x) &= iN(\bar{\psi}_p(x)\, \gamma_\mu \psi_p(x)); \\ j_\mu^\pi(x) &= -i\left[\phi^+\left(\frac{\partial \phi}{\partial x_\mu} - ieA_\mu \phi\right) - \left(\frac{\partial \phi^+}{\partial x_\mu} + ieA_\mu \phi^+\right)\phi\right]. \end{aligned}\right\} \tag{12.4}$$

The first term in (12.3) is the Hamiltonian of the interaction between the electromagnetic and proton fields ($-ej_\mu^p$ is the proton current, e is the charge of the electron); the second and third terms are the Hamiltonian of the interaction between the electromagnetic field and the field of charged π-mesons ($-ej_\mu^\pi$ is the current of charged π-mesons). The Hamiltonian of the interaction between the electromagnetic and meson fields is obtained by replacing in the free field Lagrangian (5.5) the derivatives $\partial\phi/\partial x_\mu$ and $\partial\phi^*/\partial x_\mu$ respectively by the expressions

$$\left(\frac{\partial \phi}{\partial x_\mu} - ieA_\mu \phi\right) \quad \text{and} \quad \left(\frac{\partial \phi^*}{\partial x_\mu} + ieA_\mu \phi^*\right).$$

The concrete form of the Hamiltonians $\mathcal{H}_I^{ea}$ and $\mathcal{H}_I^a$ is not required.

The S-matrix is equal to

$$S = \sum_{n=0}^{\infty} \frac{(-i)^n}{n!} \int dx_1\, dx_2 \ldots dx_n\, T(\mathcal{H}_I(x_1)\, \mathcal{H}_I(x_2) \ldots \mathcal{H}_I(x_n)). \tag{12.5}$$

The interaction Hamiltonian $\mathcal{H}_I(x)$ is given by expression (12.2). Obviously the S-matrix can also be written

$$S = T\left(e^{-i\int \mathcal{H}_I(x)\, dx}\right). \tag{12.6}$$

We are interested in the matrix element of the process of electron–nucleon scattering in the lowest order of perturbation theory for the electromagnetic interaction. In this approximation a contribution to the matrix element of the process is made only by those terms of the S-matrix in which the Hamiltonian $\mathcal{H}_I^e(x)$ enters linearly [the operator $\psi_e^{(+)}(x)$ annihilates the initial electron and $\bar{\psi}_e^{(-)}(x)$ by acting to the left annihilates the final electron]. There are no photons in the initial and final states. It is clear that a contribution in the lowest order in e is made only by those terms in which the operator $A_\nu(x)$ in $\mathcal{H}_I^e(x)$ is paired with the operator $A_\mu(y)$ in the Hamiltonian $\mathcal{H}_I^{ea}(y)$. This means that in the approximation under discussion a contribution to the matrix element of the process is made by those terms of the S-matrix in which the operator $\mathcal{H}_I^{ea}(y)$ enters linearly. Inserting (12.2) into expression (12.6) and limiting ourselves by the terms linear in $\mathcal{H}_I^e$ and $\mathcal{H}_I^{ea}$ we obtain

$$\begin{aligned} S &= T\left(e^{-i\int \mathcal{H}_I^e(x)\, dx}\, e^{-i\int \mathcal{H}_I^{ea}(y)\, dy}\, e^{-i\int \mathcal{H}_I^a(z)\, dz}\right) \\ &= T\left(\left[1+(-i)\int \mathcal{H}_I^e(x)\, dx + \ldots\right]\left[1+(-i)\int \mathcal{H}_I^{ea}(y)\, dy + \ldots\right] e^{-i\int \mathcal{H}_I^a(z)\, dz}\right) \\ &= 1+(-i)^2 T\left(\int \mathcal{H}_I^e(x)\, \mathcal{H}_I^{ea}(y)\, dx\, dy\, e^{-i\int \mathcal{H}_I^a(z)\, dz}\right) + \ldots . \end{aligned} \tag{12.7}$$

We wrote out the only term which contributes in the order e^2 to the matrix element of the process (12.1).

Let us write the Hamiltonian $\mathcal{H}_I^{ea}(x)$ in the form

$$\mathcal{H}_I^{ea}(x) = -ej_\mu^a(x)\, A_\mu(x) + (\mathcal{H}_I^{ea}(x))'. \tag{12.8}$$

Here $(\mathcal{H}_I^{ea}(x))'$ is that part of the Hamiltonian $\mathcal{H}_I^{ea}(x)$ into which $eA(x)$ enters quadratically [see (12.3)] and $ej_\mu^a(x)$ is the total current of the strongly interacting particles with $A = 0$. Only the first term of expression (12.8) contributes to the matrix element of the process we are examining to second order in e.

Denote by S^{e^2} that part of the S-matrix which contributes to the matrix element of the process (12.1) in the lowest order of the electromagnetic interaction. We have

$$S^{e^2} = -ie^2 \int dx\, dy N(\bar{\psi}_e(x)\, \gamma_\nu \psi_e(x))\, T(A_\nu(x) A_\mu(y))\, T\left(j_\mu^a(y)\, e^{-i\int \mathcal{H}^a(z)\, dz}\right). \tag{12.9}$$

We now calculate the matrix element of the process. Let q and q' (p and p') be respectively the four-momenta of the initial and final electrons (nucleons). The initial and final state vectors are written in the forms

$$\left.\begin{aligned} \Phi_{q,p} &= c^{+}(q)\,\Phi_{p}, \\ \Phi_{q',p'} &= c^{+}(q')\Phi_{p'}, \end{aligned}\right\} \tag{12.10}$$

where Φ_p and $\Phi_{p'}$ are state vectors of the initial and final nucleons. From (12.9) and (12.10) we obtain

$$\begin{aligned} (\Phi^{+}_{q'p'}S^{e^2}\Phi_{qp}) &= -ie^2\frac{1}{(2\pi)^3}\left(\frac{m^2}{q_0q_0'}\right)^{1/2}\frac{(-i)}{(2\pi)^4}(2\pi)^4\,\bar{u}(q')\,\gamma_\mu u(q) \\ &\times\int\delta(q'-q-k)\,\frac{1}{k^2}\,\mathrm{e}^{-iky}\left(\Phi^{+}_{p'}T\left(j^a_\mu(y)\,\mathrm{e}^{-i\int\mathcal{H}^a_I(x)\,dx}\right)\Phi_p\right)dk\,dy \end{aligned} \tag{12.11}$$

(m is the mass of the electron).

Perturbation theory is not applicable to strong interactions. Therefore, we are limited by the fact that we can determine on the basis of invariance principles only the general structure of the matrix element:

$$\left(\Phi^{+}_{p'}T\left(j^a_\mu(y)\,\mathrm{e}^{-i\int\mathcal{H}^a_I(x)\,dx}\right)\Phi_p\right). \tag{12.12}$$

Before we turn to this, we shall show that the hadron matrix element (12.12) can be written in a more compact form if we use the Heisenberg representation rather than the interaction representation. Let us examine the operator

$$T\left(j^a_\mu(y)\,\mathrm{e}^{-i\int\mathcal{H}^a_I(x)\,dx}\right). \tag{12.13}$$

We have

$$\begin{aligned} T\left(j^a_\mu(y)\,\mathrm{e}^{-i\int\mathcal{H}^a_I(x)\,dx}\right) &= T\left(j^a_\mu(y)\,\mathrm{e}^{-i\int_{-\infty}^{\infty}H^a_I(x_0)\,dx_0}\right) \\ &= T\left(j^a_\mu(y)\,\mathrm{e}^{-i\int_{-\infty}^{y_0}H^a_I(x_0)\,dx_0}\,\mathrm{e}^{-i\int_{y_0}^{\infty}H^a_I(x_0)\,dx_0}\right) \\ &= T\left(\mathrm{e}^{-i\int_{y_0}^{\infty}H^a_I(x_0)\,dx_0}\right)j^a_\mu(y)\,T\left(\mathrm{e}^{-i\int_{-\infty}^{y_0}H^a_I(x_0)\,dx_0}\right). \end{aligned} \tag{12.14}$$

Here

$$H^a_I(x_0) = \int\mathcal{H}^a_I(x)\,dx. \tag{12.15}$$

Remembering the definition of the operator $U(t, t_1)$ [see (1.20) and (1.35)] we obtain

$$T\left(\mathrm{e}^{-i\int_{y_0}^{\infty}H^a_I(x_0)\,dx_0}\right) = \sum_{n=0}^{\infty}\frac{(-i)^n}{n!}\int_{y_0}^{\infty}dx_{10}\ldots\int_{y_0}^{\infty}dx_{n0}T(H^a_I(x_{10})\ldots H^a_I(x_{n0})) = U_a(\infty, y_0), \tag{12.16}$$

$$T\left(\mathrm{e}^{-i\int_{-\infty}^{y_0}H^a_I(x_0)\,dx_0}\right) = U_a(y_0, -\infty). \tag{12.17}$$

We emphasize that the operators $U_a(\infty, y_0)$ and $U_a(y_0, -\infty)$ are determined by the strong interaction Hamiltonian $\mathcal{H}^a_I$. For example, the operator $U_a(y_0, -\infty)$ acting on the state vector given for $t \to -\infty$ converts it into the state vector at time y_0 *due to the strong interactions*. From (12.14), (12.16), and (12.17) we find

$$T(j^a_\mu(y)\, e^{-i\int \mathcal{H}^a_I(x)\, dx}) = U_n(\infty, y_0)\, j^a_\mu(y)\, U_a(y_0, -\infty). \tag{12.18}$$

Let us examine a system in which only strong interactions take place. We have used the interaction representation in which the state vector $\Phi(t)$ of such a system satisfies the equation

$$i\frac{\partial \Phi(t)}{\partial t} = H^a_I(t)\,\Phi(t), \tag{12.19}$$

and any operator $O(t)$ belonging to the system obeys the equation

$$i\frac{\partial O(t)}{\partial t} = [O(t), H^a_0] \tag{12.20}$$

(H^a_0 is the free Hamiltonian).

We perform the unitary transformation from $\Phi(t)$ and $O(t)$ to

$$\left.\begin{aligned} \Phi_\Gamma(t) &= U^+_a(t, t_0)\,\Phi(t), \\ O_\Gamma(t) &= U^+_a(t, t_0)\, O(t)\, U_a(t, t_0). \end{aligned}\right\} \tag{12.21}$$

Here t_0 is an arbitrary fixed moment of time and $U_a(t, t_0)$ is the unitary operator defined by expression (1.35). Obviously, for $t = t_0$

$$\Phi_\Gamma(t_0) = \Phi(t_0), \qquad O_\Gamma(t_0) = O(t_0). \tag{12.22}$$

The operators $U_a(t, t_0)$ and $U^+_a(t, t_0)$ satisfy the equations [see (1.21) and (1.22)]:

$$i\frac{\partial U_a(t, t_0)}{\partial t} = H^a_I(t)\, U_a(t, t_0), \tag{12.23a}$$

$$-i\frac{\partial U^+_a(t, t_0)}{\partial t} = U^+_a(t, t_0)\, H^a_I(t). \tag{12.23b}$$

As was shown in the first section, observable quantities are invariant under unitary transformations (12.21). We shall obtain the equation of motion for the state vector $\Phi_\Gamma(t)$. From (12.19), (12.21), and (12.23) we find that

$$\frac{\partial \Phi_\Gamma(t)}{\partial t} = 0. \tag{12.24}$$

Thus, the vector Φ_Γ which describes the same state as the vector $\Phi(t)$ in the interaction representation and which is related to $\Phi(t)$ by relationship (12.21), does not depend on time. The representation in which state vectors do not depend on time is called the *Heisenberg representation*. All dependence on time in this representation is transferred to the operators.

With the help of (12.21) and (12.23) we find

$$i\frac{\partial O_\Gamma(t)}{\partial t} = U^+_a(t, t_0)([O(t), H^a_I(t)] + [O(t), H^a_0])\, U_a(t, t_0) = [O_\Gamma(t)\, H^a_\Gamma(t)]. \tag{12.25}$$

Here

$$H_T^a(t) = U_a^+(t, t_0)\big(H_I^a(t)+H_0^a\big)\, U_a(t, t_0)$$

is the operator of the total energy in the Heisenberg representation. For $H_T^a(t)$, we obtain from (12.25)

$$\frac{\partial H_T^a(t)}{\partial t} = 0. \tag{12.26}$$

The operator of the total energy in the Heisenberg representation therefore does not depend on time.

We perform the transition from the interaction representation to the Heisenberg representation in the matrix element (12.12). Suppose $t_0 \to -\infty$ [this means that both representations coincide, see (12.22)]. In the matrix element (12.12) Φ_p is a state vector in the interaction representation for $t \to -\infty$ and $\Phi_{p'}$ is a state vector for $t \to \infty$. From (12.22) it follows that the vector Φ_p coincides with the corresponding vector in the Heisenberg representation. Denote it by $\Phi_{p;\,\mathrm{in}}$. The vector in the Heisenberg representation corresponding to $\Phi_{p'}$ is equal to

$$\Phi_{p';\,\mathrm{out}} = U_a^+(\infty, -\infty)\Phi_{p'} = S_a^+\Phi_{p'}. \tag{12.27}$$

Using the unitarity of the operator $U_a(t, t_0)$ [see (1.25)] we obtain from (12.18)

$$\begin{aligned}&\big(\Phi_{p'}^+ T\big(j_\mu^a(y)\, \mathrm{e}^{-i\int \mathscr{H}_I^a(x)\,dx}\big)\Phi_p\big)\\ &= (\Phi_{p'}^+ U_a(\infty, y_0)\, U_a(y_0, -\infty)\, U_a^+(y_0, -\infty)\, j_\mu^a(y)\, U_a(y_0, -\infty)\Phi_p)\\ &= (\Phi_{p';\,\mathrm{out}}^+ J_\mu^a(y)\, \Phi_{p;\,\mathrm{in}}).\end{aligned} \tag{12.28}$$

Here

$$J_\mu^a(y) = U_a^+(y_0, -\infty) j_\mu^a(y)\, U_a(y_0, -\infty) \tag{12.29}$$

is the current operator in the Heisenberg representation. Note that in obtaining (12.28) we used the following obvious relationship:

$$U_a(\infty, y_0)\, U_a(y_0, -\infty) = S_a.$$

Thus, the matrix element (12.12) is the matrix element of the current in the Heisenberg representation. The vector $\Phi_{p'}$ describes a nucleon with four-momentum p'. In acting on such a vector with the matrix S_a we obviously obtain the vector $\Phi_{p'}$ (the nucleon cannot be transformed by strong interactions into any other particles). Thus, the vector $\Phi_{p';\,\mathrm{out}}$ coincides with the vector $\Phi_{p'}$ in the interaction representation up to a phase factor. Henceforth, the indices in and out will, as a rule, be omitted.

Any operator in the Heisenberg representation satisfies the relationship

$$i\,\frac{\partial F(y)}{\partial y_\alpha} = [P_\alpha^a, F(y)], \tag{12.30}$$

where P^a_α is the total four-momentum operator [we proved this relationship for $\alpha = 4$; see (12.25)]. From (12.30) we find†

$$i\frac{\partial}{\partial y_\alpha}(\Phi^+_{p'}J^a_\mu(y)\Phi_p) = (\Phi^+_{p'}[P^a_\alpha, J^a_\mu(y)]\Phi_p) = (p'-p)_\alpha(\Phi^+_{p'}J^a_\mu(y)\Phi_p). \tag{12.31}$$

Hence we have

$$(\Phi^+_{p'}J^a_\mu(y)\Phi_p) = e^{-i(p'-p)y}(\Phi^+_{p'}J^a_\mu(0)\Phi_p). \tag{12.32}$$

Inserting (12.32) into (12.11) we obtain the following expression for the matrix element for electron–nucleon scattering:

$$(\Phi^+_{q'p'}S^{e^2}\Phi_{qp}) = -\frac{e^2}{(2\pi)^3}\left(\frac{m^2}{q_0q'_0}\right)^{1/2}(2\pi)^4\,\bar{u}(q')\gamma_\mu u(q)$$
$$\times\frac{1}{(q'-q)^2}(\Phi^+_{p'}J^a_\mu(0)\Phi_p)\,\delta(q'+p'-q-p). \tag{12.33}$$

If the strong interactions of the nucleon are not considered, then obviously the matrix element $(\Phi^+_{p'}J^a_\mu(0)\Phi_p)$ for electron–proton scattering is equal to‡

$$-\frac{i}{(2\pi)^3}\left(\frac{M^2}{p_0p'_0}\right)^{1/2}\bar{u}(p')\gamma_\mu u(p)$$

(M is the mass of the nucleon). In this case the scattering process is described by the diagram in Fig. 17 (the double lines relate to the nucleon).

In the general case the nucleon matrix element can be written in the form

$$(\Phi^+_{p'}J^a_\mu(0)\Phi_p) = -\frac{i}{(2\pi)^3}\left(\frac{M^2}{p_0p'_0}\right)^{1/2}\bar{u}(p')\,\Gamma_\mu(p',p)\,u(p). \tag{12.34}$$

Here $\Gamma_\mu(p',p)$ is a matrix acting on the spin variables. Inserting (12.34) into (12.33) we obtain for the matrix element

$$(\Phi^+_{q',p'}S^{e^2}\Phi_{q,p}) = i\frac{e^2}{(2\pi)^2}\left(\frac{m^2M^2}{q_0q'_0p_0p'_0}\right)^{1/2}\bar{u}(q')\gamma_\mu u(q)$$
$$\times\frac{1}{(q'-q)^2}\bar{u}(p')\,\Gamma_\mu(p',p)\,u(p)\,\delta(p'+q'-p-q). \tag{12.35}$$

We put the matrix element (12.35) into correspondence with the diagram depicted in Fig. 18. The hatched part of the diagram is determined by strong interactions. The matrix $\Gamma_\mu(p',p)$ corresponds to this element of the diagram.

† The states Φ_p and $\Phi_{p'}$ are eigenstates of the energy-momentum operator of the free nucleon field. In obtaining (12.31) the adiabatic hypothesis was used, according to which the interaction is absent when $t \to \pm\infty$.

‡ Remember that the electric current is equal to eJ^a_μ, where e is the charge of the electron.

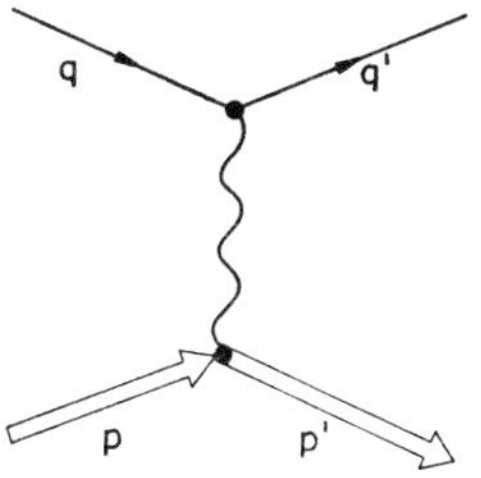

FIG. 17

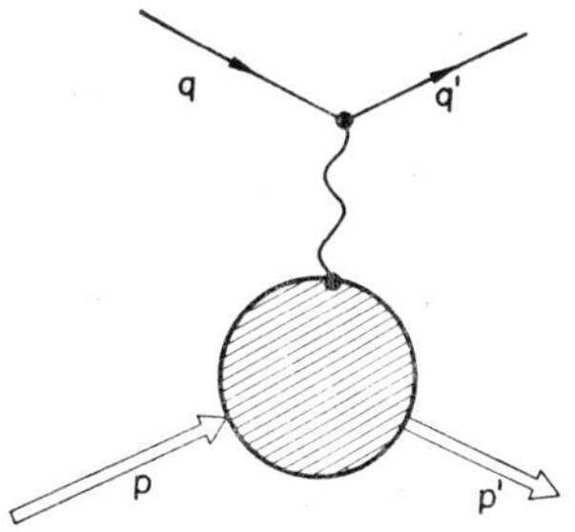

FIG. 18

Let us ascertain the general structure of the matrix $\Gamma_\mu(p', p)$. Under Lorentz transformations the matrix element

$$\bar{u}(p')\,\Gamma_\mu(p', p)\,u(p)$$

must transform the same way as $\bar{u}(p')\gamma_\mu u(p)$, i.e. as a vector. We expand the 4×4 matrix $\Gamma_\mu(p', p)$ in the full system of sixteen Dirac matrices. We obtain

$$\bar{u}(p')\,\Gamma_\mu(p', p)\,u(p) = \bar{u}(p')\,[A_\mu + B_{\mu\nu}\gamma_\nu + C_{\mu\nu\varrho}\sigma_{\nu\varrho} + D_{\mu\nu}\gamma_\nu\gamma_5 + E_\mu\gamma_5]\,u(p), \tag{12.36}$$

where

$$\sigma_{\nu\varrho} = \frac{1}{2i}(\gamma_\nu\gamma_\varrho - \gamma_\varrho\gamma_\nu).$$

The coefficients A_μ, $B_{\mu\nu}$ . . . of the expansion depend on p and p'.

The first terms A_μ of the expansion must be a vector. We obtain the following general expression for A_μ:

$$A_\mu = a_1 p_\mu + a_2 p'_\mu. \tag{12.37}$$

Here a_1 and a_2 are functions of scalars which can be constructed from the vectors p and p'. Since $p^2 = -M^2$ and $(p')^2 = -M^2$, the only scalar which can be constructed from p and p' is pp'.

Define the momentum transfer four-vector $\varkappa$ by

$$\varkappa = p' - p. \tag{12.38}$$

Obviously, $pp' = -\varkappa^2/2 - M^2$. Thus, a_1 and a_2 are quadratic functions of the momentum transfer $\varkappa^2$.

We now turn to the construction of the succeeding terms in the expansion (12.36). Since the matrix element $\bar{u}(p')\gamma_\nu u(p)$ transforms like a vector, the quantities $B_{\mu\nu}$ must transform like a second-rank tensor. We obtain the following general expression for the tensor $B_{\mu\nu}$:

$$B_{\mu\nu} = b_1 p_\mu p_\nu + b_2 p'_\mu p'_\nu + b_3 p'_\mu p_\nu + b_4 p_\mu p'_\nu + b\delta_{\mu\nu}, \tag{12.39}$$

where b, b_1, . . . are functions of $\varkappa^2$. The spinors $u(p)$ and $\bar{u}(p')$ satisfy Dirac's equation (see Appendix)

$$(\hat{p} - iM)\,u(p) = 0, \tag{12.40a}$$

$$\bar{u}(p')\,(\hat{p}' - iM) = 0. \tag{12.40b}$$

With the help of (12.40) it is not difficult to show that the first four terms of (12.39) after insertion into (12.36) reduce to (12.37). Actually,

$$\bar{u}(p')p_\mu \hat{p}u(p) = \bar{u}(p')iMp_\mu u(p), \quad \bar{u}(p')p'_\mu \hat{p}'u(p) = \bar{u}(p')iMp'_\mu u(p)$$

and so forth. As a result we conclude that we need to consider only the term

$$b\delta_{\mu\nu}. \tag{12.41}$$

The quantities $C_{\mu\nu\varrho}$ must transform like third-rank tensors. This tensor must be antisymmetric under the transposition of the indices ν and ϱ ($\sigma_{\nu\varrho} = -\sigma_{\varrho\nu}$). The general expression for $C_{\mu\nu\varrho}$ has the form

$$C_{\mu\nu\varrho} = c_1 p_\mu(p_\nu p'_\varrho - p_\varrho p'_\nu) + c_2 p'_\mu(p_\nu p'_\varrho - p_\varrho p'_\nu) + c_3(\delta_{\mu\nu}p_\varrho - \delta_{\mu\varrho}p_\nu) + c_4(\delta_{\mu\nu}p'_\varrho - \delta_{\mu\varrho}p'_\nu).$$

It is not difficult to see that all terms of this tensor after insertion into (12.36) reduce to (12.37) and (12.41). Actually, using equations (12.40) we find

$$\begin{aligned} \bar{u}(p')\,p_\mu\sigma_{\nu\varrho}(p_\nu p'_\varrho - p_\varrho p'_\nu)\,u(p) &= \bar{u}(p')\,i\varkappa^2 p_\mu u(p), \\ \bar{u}(p')\,\sigma_{\nu\varrho}(\delta_{\mu\nu}p_\varrho - \delta_{\mu\varrho}p_\nu)\,u(p) &= 2\bar{u}(p')\,[M\gamma_\mu + ip_\mu]\,u(p) \end{aligned}$$

and so forth.

The quantities $D_{\mu\nu}$ must transform like a pseudotensor of the second rank. We obtain the general expression

$$D_{\mu\nu} = \varepsilon_{\mu\nu\varrho\sigma}p_\varrho p'_\sigma d$$

($\varepsilon_{\mu\nu\varrho\sigma}$ is antisymmetric under the transposition of any tensor indices; $\varepsilon_{\mu\nu\varrho\sigma} = -\varepsilon_{\mu\nu\varrho\sigma} = \ldots$; $\varepsilon_{1234} = 1$). This term after being inserted into (12.36) also reduces to (12.37) and (12.41). To show this we use the relationship

$$\varepsilon_{\mu\nu\varrho\sigma}\gamma_\nu\gamma_5 = \gamma_\sigma\gamma_\mu\gamma_\varrho + \gamma_\mu\delta_{\varrho\sigma} - \gamma_\varrho\delta_{\mu\sigma} + \gamma_\sigma\delta_{\mu\varrho}, \tag{12.42}$$

whose validity can be verified easily. Using (12.40) and (12.42) we obtain

$$\bar{u}(p')\,\varepsilon_{\mu\nu\varrho\sigma}\gamma_\nu\gamma_5\,u(p)\,p_\varrho p'_\sigma = -\bar{u}(p')\,[(M^2 - pp')\,\gamma_\mu + iM(p'_\mu + p_\mu)]\,u(p). \tag{12.43}$$

Thus, the term $D_{\mu\nu}$ also need not be considered. The last term E_μ must be a pseudovector. Since it is impossible to construct a pseudovector from p_μ and p'_μ, $E_\mu = 0$.

Thus, if we require that the matrix element

$$\bar{u}(p')\,\Gamma_\mu(p', p)\,u(p)$$

transform like a vector, we conclude that the matrix $\Gamma_\mu(p', p)$ can be written in the form of a sum

$$a\gamma_\mu + bp_\mu + cp'_\mu.$$

Let us write this matrix in another form. We express p and p' by the vectors P and $\varkappa$ which are defined by

$$P = p + p', \quad \varkappa = p' - p. \tag{12.44}$$

Considering the relationships

$$\left.\begin{aligned}\bar{u}(p')\,\sigma_{\mu\nu}\varkappa_\nu u(p) &= \frac{1}{i}\,\bar{u}(p')\,[\gamma_\mu(\hat{p}'-\hat{p})-\varkappa_\mu]\,u(p)\\ &= -\bar{u}(p')\,[2M\gamma_\mu+iP_\mu]\,u(p),\\ \bar{u}(p')\,\sigma_{\mu\nu}P_\nu u(p) &= (-i)\,\bar{u}(p')\,u(p)\,\varkappa_\mu,\end{aligned}\right\} \tag{12.45}$$

we write the matrix $\Gamma_\mu(p', p)$ in the form

$$\Gamma_\mu(p', p) = F_1(\varkappa^2)\,\gamma_\mu - \frac{F_2(\varkappa^2)}{2M}\,\sigma_{\mu\nu}\,\varkappa_\nu - \frac{F_3(\varkappa^2)}{2M}\,\sigma_{\mu\nu}P_\nu. \tag{12.46}$$

We now examine the limitations imposed on the matrix $\Gamma_\mu(p', p)$ by the law of conservation of current

$$\frac{\partial J_\mu^a(y)}{\partial y_\mu} = 0.$$

With the help of (12.32) we obtain

$$\frac{\partial}{\partial y_\mu}\,(\Phi_{p'}^+ J_\mu(y)\,\Phi_p) = (-i)\,e^{-i(p'-p)y}\,(p'-p)_\mu\,(\Phi_{p'}^+ J_\mu^a(0)\Phi_p) = 0.$$

Thus, it follows from current conservation that

$$(p'-p)_\mu\,\bar{u}(p')\,\Gamma_\mu(p', p)\,u(p) = 0. \tag{12.47}$$

Considering (12.40) and (12.45) we find

$$(p'-p)_\mu\,\bar{u}(p')\,\Gamma_\mu(p', p)\,u(p) = \frac{i}{2M}\,F_3\varkappa^2\,\bar{u}(p')\,u(p).$$

Hence, $F_3 = 0$. The matrix $\Gamma_\mu(p', p)$ consequently is equal to

$$\Gamma_\mu(p', p) = F_1(\varkappa^2)\,\gamma_\mu - \frac{F_2(\varkappa^2)}{2M}\,\sigma_{\mu\nu}\varkappa_\nu. \tag{12.48}$$

Thus, we have shown that the matrix element $(\Phi_{p'}^+ J_\mu^a(0)\Phi_p)$ (the hatched part of the diagram in Fig. 18) is characterized by two quadratic functions of the four-momentum transfer. These functions are called *electromagnetic form factors* of the nucleon (F_1 is the Dirac form factor and F_2 is the Pauli form factor).

We shall show that the form factors F_1 and F_2 are real. To do this we use the condition of unitarity of the S-matrix [see (1.39)]:

$$S^+S = 1.$$

Write the S-matrix in the form

$$S = 1+R. \tag{12.49}$$

From the condition of unitarity it follows that

$$R+R^+ = -R^+R. \tag{12.50}$$

For the matrix element of the process of electron–nucleon scattering we obtain

$$(\Phi^+_{q',p'}R\Phi_{q,p})+(\Phi^+_{q',p'}R^+\Phi_{q,p}) = -\sum_n (\Phi^+_{q',p'}R^+\Phi_n)(\Phi^+_n R\Phi_{q,p}). \tag{12.51}$$

On the right-hand side of this relationship there is the sum over all the states which the electron–nucleon system can change into. It is obvious that in any state Φ_n there must be at least one electron and one baryon. This means that the right-hand side of (12.51) is of order e^4. Thus, in the e^2-approximation we have been examining we obtain from the unitarity relationship

$$(\Phi^+_{q',p'}R\Phi_{q,p}) = -(\Phi^+_{q,p}R\Phi_{q',p'})^*. \tag{12.52}$$

Inserting expression (12.35) into (12.52) we find

$$(\bar{u}(q')\gamma_\mu u(q))(\bar{u}(p')\Gamma_\mu(p',p)u(p)) = (\bar{u}(q)\gamma_\mu u(q'))^*(\bar{u}(p)\Gamma_\mu(p,p')u(p'))^*. \tag{12.53}$$

It is not difficult to see that

$$(\bar{u}(q)\gamma_\mu u(q'))^* = (u^+(q')\gamma^+_\mu\gamma_4 u(q)) = (\bar{u}(q')\bar{\gamma}_\mu u(q)), \tag{12.54}$$

where

$$\bar{\gamma}_\mu = \gamma_4\gamma^+_\mu\gamma_4.$$

Since $\gamma^+_\mu = \gamma_\mu$, with the help of the commutation relations for the matrices γ_μ we obtain

$$\bar{\gamma}_\mu = \gamma_\mu\delta_\mu, \tag{12.55}$$

where

$$\delta_\mu = \begin{cases} -1, & \mu = 1, 2, 3; \\ 1, & \mu = 4 \end{cases} \tag{12.56}$$

[in (12.55) there is to be no summation over μ].

From (12.53)–(12.55) we find

$$(\bar{u}(p')\Gamma_\mu(p',p)u(p)) = \delta_\mu(\bar{u}(p)\Gamma_\mu(p,p')u(p'))^*. \tag{12.57}$$

Inserting expression (12.48) into (12.57) and recognizing the fact that

$$(\bar{u}(p)\sigma_{\mu\nu}\varkappa_\nu u(p'))^*\delta_\mu = (\bar{u}(p')\gamma_4\sigma_{\mu\nu}\gamma_4\varkappa^*_\nu u(p))\delta_\mu = -(\bar{u}(p')\sigma_{\mu\nu}\varkappa_\nu u(p)),$$

we obtain

$$F^*_1(\varkappa^2) = F_1(\varkappa^2), \quad F^*_2(\varkappa^2) = F_2(\varkappa^2). \tag{12.58}$$

Thus, the hatched part of the diagram in Fig. 18 is characterized by two real quadratic functions of the four-momentum transfer.

Instead of the form factors F_1 and F_2 the magnetic and charge form factors G_M and G_E are often introduced. They are related to F_1 and F_2 by the relationships

$$G_M = F_1+F_2, \quad G_E = F_1-\frac{\varkappa^2}{4M^2}F_2. \tag{12.59}$$

Let us clarify the physical meaning of these form factors. With the help of (12.45) the matrix element of the current can be written

$$(\Phi_{p'}^{+}J_{\mu}^{a}(0)\,\Phi_{p}) = -i\frac{1}{(2\pi)^3}\left(\frac{M^2}{p_0 p_0'}\right)^{1/2}\bar{u}(p')\left[\gamma_\mu(F_1+F_2)+iP_\mu\frac{1}{2M}F_2\right]u(p). \tag{12.60}$$

Let us examine this expression in the system in which $\boldsymbol{p}+\boldsymbol{p}' = 0$ (the Breit frame). Considering the fact that in this system $p_0 = p_0'$ we obtain

$$\left.\begin{aligned}(\Phi_{p'}^{+}\boldsymbol{J}^{a}(0)\,\Phi_{p}) &= -i\frac{1}{(2\pi)^3}\left(\frac{M}{p_0}\right)(\bar{u}(p')\,\boldsymbol{\gamma}u(p))(F_1+F_2),\\ (\Phi_{p'}^{+}J_{4}^{a}(0)\,\Phi_{p}) &= -i\frac{1}{(2\pi)^3}\left(\frac{M}{p_0}\right)\bar{u}(p')\left[\gamma_4(F_1+F_2)-\frac{p_0}{M}F_2\right]u(p).\end{aligned}\right\} \tag{12.61}$$

From equations (12.40) it is not difficult to see that

$$\bar{u}(p')(\hat{p}'+\hat{p})\,u(p) = 2iM\bar{u}(p')\,u(p), \tag{12.62}$$

from which in the Breit frame

$$p_0\bar{u}(p')\,\gamma_4 u(p) = M\bar{u}(p')\,u(p). \tag{12.63}$$

Using this relationship and considering the fact that in the Breit frame $\varkappa^2 = 4\boldsymbol{p}^2$ we obtain

$$(\Phi_{p'}^{+}J_{4}^{a}(0)\,\Phi_{p}) = -i\frac{1}{(2\pi)^3}\left(\frac{M}{p_0}\right)(\bar{u}(p')\,\gamma_4 u(p))\left(F_1-\frac{\varkappa^2}{4M^2}F_2\right). \tag{12.64}$$

The spinors $u(p)$ and $u(p')$ in the Dirac–Pauli representation have the form (see Appendix)

$$\left.\begin{aligned}u(p) &= \left(\frac{p_0+M}{2M}\right)^{1/2}\begin{pmatrix}\phi\\ \dfrac{\boldsymbol{\sigma p}}{p_0+M}\,\phi\end{pmatrix},\\ u(p') &= \left(\frac{p_0+M}{2M}\right)^{1/2}\begin{pmatrix}\phi'\\ \dfrac{-\boldsymbol{\sigma p}}{p_0+M}\,\phi'\end{pmatrix}\end{aligned}\right\} \tag{12.65}$$

(the system $\boldsymbol{p}+\boldsymbol{p}' = 0$). With the help of these expressions it is not difficult to see that

$$\left.\begin{aligned}\bar{u}(p')\,\gamma_4 u(p) &= \phi'^{+}\phi,\\ \bar{u}(p')\,\boldsymbol{\gamma}u(p) &= \phi'^{+}\frac{\boldsymbol{\sigma}\times\boldsymbol{\varkappa}}{2M}\,\phi.\end{aligned}\right\} \tag{12.66}$$

From (12.61), (12.64), and (12.66) we obtain the following expressions for the matrix elements of the operators $\boldsymbol{J}^a(0)$ and $J_4^a(0)$ in the Breit frame:

$$\left.\begin{aligned}(\Phi_{p'}^{+}\boldsymbol{J}^{a}(0)\,\Phi_{p}) &= -i\frac{1}{(2\pi)^3}\left(\frac{M}{p_0}\right)\left(\phi'^{+}\frac{\boldsymbol{\sigma}\times\boldsymbol{\varkappa}}{2M}\,\phi\right)G_M(\varkappa^2),\\ (\Phi_{p'}^{+}J_{4}^{a}(0)\,\Phi_{p}) &= -i\frac{1}{(2\pi)^3}\left(\frac{M}{p_0}\right)(\phi'^{+}\phi)\,G_E(\varkappa^2).\end{aligned}\right\} \tag{12.67}$$

The form factor $G_M(\varkappa^2)$ characterizes thus the distribution of the magnetic moment of the nucleon and $G_E(\varkappa^2)$ the distribution of the charge. For $\varkappa^2 = 0$ the form factors of the proton are†

$$\left.\begin{aligned} G_M(0) &= F_1(0)+F_2(0) = \mu_p, \\ G_E(0) &= F_1(0) = 1, \end{aligned}\right\} \qquad (12.68)$$

where μ_p is the full magnetic moment of the proton (in nuclear magnetons). It follows from this that $F_2(0) = \mu_p - 1$, i.e. $F_2(0)$ is equal to the anomalous magnetic moment of the proton (in nuclear magnetons). For the neutron

$$\left.\begin{aligned} G_M(0) &= F_1(0)+F_2(0) = \mu_n, \\ G_E(0) &= F_1(0) = 0. \end{aligned}\right\} \qquad (12.69)$$

Here μ_n is the magnetic moment of the neutron in nuclear magnetons.

With this we conclude our examination of the process of electron–nucleon scattering. The cross section for this process will be calculated in § 14.

We shall now examine electron scattering on a particle with spin zero (for example, π–e-scattering). Denote by q and q' the four-momenta of the initial and final electrons respectively and by p and p' the initial and final four-momenta of the particles with spin zero. It is obvious that in the second order of perturbation theory in e the matrix element of the process is given by expression (12.33) in which Φ_p and $\Phi_{p'}$ are the initial and final state vectors of the particle with spin zero. If one does not consider strong interactions, the matrix element of current is

$$(-i)\left(\Phi_{p'}^{+} N\left(\Phi^{+}(x)\frac{\partial\phi(x)}{\partial x_\mu} - \frac{\partial\phi^{+}(x)}{\partial x_\mu}\phi(x)\right)\Phi_p\right) = \frac{1}{(2\pi)^3}\frac{1}{\sqrt{4p_0p_0'}}\, e^{-i(p'-p)x}(p_\mu+p_\mu'). \qquad (12.70)$$

In the general case the matrix element of current can be written in the form

$$(\Phi_{p'}^{+} J_\mu^a(0)\Phi_p) = \frac{1}{(2\pi)^3}\frac{1}{\sqrt{4p_0p_0'}}\Lambda_\mu(p', p), \qquad (12.71)$$

† Actually, in the Breit frame

$$-ie(\Phi_{p'}^{+} \int J_4^a(x)\, dx\, \Phi_p) = -ie(2\pi)^3\,\delta(\varkappa)\,(\Phi_{p'}^{+} J_4(0)\,\Phi_p) = -e\left(\frac{M}{p_0}\right) G_E(\varkappa^2)\,\delta(\varkappa)$$

e is the charge of the electron; we assumed $\phi' = \phi$). On the other side

$$-ie(\Phi_{p'}^{+} \int J_4^a(x)\, dx\, \Phi_p) = e_N(\Phi_{p'}^{+}\Phi_p) = e_N\,\delta(\varkappa),$$

where e_N is the charge of the nucleon. Integrating over $\varkappa$ we obtain $G_E(0) = e_N/(-e)$. Consider the matrix element of the magnetic moment operator

$$\mathfrak{M} = e\tfrac{1}{2}\int \boldsymbol{x}\times \boldsymbol{J}^a(x)\, dx.$$

We obtain

$$(\Phi_{p'}^{+}\mathfrak{M}_i\Phi_p) = \tfrac{1}{2} ie\varepsilon_{ikl}(2\pi)^3 < p' \mid J_l^a(0) \mid p > \frac{\partial\delta(\varkappa)}{\partial\varkappa_k}.$$

Integrating over $\varkappa$ we find with the help of (12.67)

$$\int (\Phi_{p'}^{+}\mathfrak{M}_i\Phi_p)\, d\varkappa = \frac{(-e)}{2M} G_M(0)\, \phi'^{+}\sigma_i\phi.$$

Hence $G_M(0)$ is the magnetic moment of the nucleon in nuclear magnetons.

where the quantity $\Lambda_\mu(p', p)$ depends on the momenta p and p' and transforms like a four-vector. Obviously $\Lambda_\mu(p', p)$ has the form

$$\Lambda_\mu(p', p) = F(\varkappa^2)\, P_\mu + F'(\varkappa^2)\, \varkappa_\mu, \tag{12.72}$$

where $P = p+p'$; $\varkappa = p'-p$.

From the law of current conservation we obtain

$$\varkappa_\mu \Lambda_\mu(p', p) = 0. \tag{12.73}$$

Inserting expression (12.72) into (12.73) we find

$$\varkappa^2 F'(\varkappa^2) = 0.$$

From this $F'(\varkappa^2) = 0$ and

$$\Lambda_\mu(p', p) = P_\mu F(\varkappa^2). \tag{12.74}$$

Finally, from the unitarity of the S-matrix in the second order of perturbation theory in e we obtain

$$\Lambda_\mu(p', p) = -\delta_{\mu}\Lambda_\mu^*(p, p'). \tag{12.75}$$

From (12.74) and (12.75) we have

$$F^*(\varkappa^2) = F(\varkappa^2). \tag{12.76}$$

Ultimately,

$$(\Phi_{p'}^+ J_\mu^a(0) \Phi_p) = \frac{1}{(2\pi)^3} \frac{1}{\sqrt{4p_0 p_0'}} (p+p')_\mu F(\varkappa^2). \tag{12.77}$$

Thus, the matrix element of the current in the case of a particle with spin zero is characterized by one real form factor. Note that the matrix element of electron scattering on a particle with spin zero in the second order of perturbation theory for the electromagnetic interaction can be represented by the diagram in Fig. 18. The hatched part of the diagram is to be put into correspondence with

$$eF(\varkappa^2)\,(p_\mu + p_\mu')\,(2\pi)^4\,\delta(p'-p+k),$$

determined by strong interactions, and the external lines with momenta p and p', respectively,

$$\frac{1}{(2\pi)^{3/2}} \frac{1}{\sqrt{2p_0}} \quad \text{and} \quad \frac{1}{(2\pi)^{3/2}} \frac{1}{\sqrt{2p_0'}}.$$

We now turn to an examination of the following processes with weak interaction:

$$\nu_\mu + n \to \mu^- + p, \quad (12.78\text{a}) \qquad \bar{\nu}_\mu + p \to \mu^+ + n. \quad (12.78\text{b})$$

Perturbation theory is applicable to weak interactions. We shall treat strong interactions phenomenologically by introducing form factors.

At the present time it is considered established that the Hamiltonian describing the weak interaction of nucleons, μ-mesons, and neutrons has the form

$$\mathcal{H}_1^w(x) = \frac{G}{\sqrt{2}}\left(\bar{\psi}_\mu(x)\,\gamma_\alpha(1+\gamma_5)\,\psi_\nu(x)\right)\left(v_\alpha(x)+a_\alpha(x)\right)$$
$$+\frac{G}{\sqrt{2}}\left(\bar{\psi}_\nu(x)\,\gamma_\alpha(1+\gamma_5)\,\psi_\mu(x)\right)\left(v_\alpha^+(x)+a_\alpha^+(x)\right)\delta_\alpha. \quad (12.79)$$

Here G is the weak interaction constant ($G \simeq 10^{-5}/M$, M is the mass of the nucleon); ψ_μ and ψ_ν are the μ-meson and neutron field operators;

$$\left.\begin{aligned} v_\alpha(x) &= \bar{\psi}_p(x)\,\gamma_\alpha\psi_n(x) - i\sqrt{2}\left(\phi(x)\,\frac{\partial\phi_0(x)}{\partial x_\alpha} - \phi_0(x)\,\frac{\partial\phi(x)}{\partial x_\alpha}\right) + \ldots, \\ a_\alpha(x) &= \bar{\psi}_p(x)\,\gamma_\alpha\gamma_5\psi_n(x) + \ldots \end{aligned}\right\} \quad (12.80)$$

are the vector and axial currents. In (12.80) ϕ is the field operator for the charged π-meson and ϕ_0 is the field operator for the neutral π-mesons. The vector current $v_\alpha(x)$ is written in accordance with the hypothesis of conserved vector current.

In examining the processes (12.78) we shall consider weak and strong interactions. Accordingly, the interaction Hamiltonian is

$$\mathcal{H}_I(x) = \mathcal{H}_I^w(x) + \mathcal{H}_I^q(x). \quad (12.81)$$

We begin with an examination of the process (12.78a). Denote by q and q' the four-momenta of the neutrino and μ-meson and by p and p' the four-momenta of the neutron and the proton. The initial and final state vectors are written in the forms

$$\Phi_{q,p} = c_\nu^+(q)\Phi_p, \quad \Phi_{q',p'} = c_\mu^+(q')\Phi_{p'}, \quad (12.82)$$

where Φ_p and $\Phi_{p'}$ are vectors describing the initial neutron and final proton. We isolate that part of the S-matrix which makes a non-zero contribution to the matrix element of the process (12.78a) to first order in the weak interaction constant G. Denote it by S^G. By the same argument as in the discussion of electron–nucleon scattering we obtain

$$S^G = (-i)\frac{G}{\sqrt{2}}\int\left(\bar{\psi}_\mu(x)\,\gamma_\alpha(1+\gamma_5)\,\psi_\nu(x)\right)\left[T\left(v_\alpha(x)+a_\alpha(x)\right)e^{-i\int \mathcal{H}_I^q(y)\,dy}\right]dx. \quad (12.83)$$

The mass of the neutrino is assumed to be zero. The spinor $u(q)$ describing the neutrino with four-momentum q will be normalized with the condition $u^+(q)u(q) = 1$ (see Appendix). With this normalization the operator $\psi_\nu(x)$ is equal to

$$\psi_\nu(x) = \frac{1}{(2\pi)^{3/2}}\int\left(u^r(q)\,c_r(q)\,e^{iqx} + u^r(-q)\,d_r^+(q)\,e^{-iqx}\right)d\boldsymbol{q}. \quad (12.84)$$

With the help of (12.83) and (12.84) we obtain the following expression for the matrix element of the process (12.78a):

$$(\Phi_{q',p'}^+ S^G \Phi_{q,p}) = (-i)\frac{1}{(2\pi^3)}\left(\frac{m}{q_0'}\right)^{1/2}\frac{G}{\sqrt{2}}\,\bar{u}(q')\,\gamma_\alpha(1+\gamma_5)\,u(q)$$
$$\times\int e^{-i(q'-q)x}\left(\Phi_{p'}^+ T\left(\left(v_\alpha(x)+a_\alpha(x)\right)e^{-i\int \mathcal{H}_I^q(y)\,dy}\right)\Phi_p\right)dx, \quad (12.85)$$

where m is the mass of the μ-meson. Transforming from the interaction representation to the Heisenberg representation we obtain

$$\left.\begin{aligned}\left(\Phi_{p'}^{+}T\left(v_{\alpha}(x)\,e^{-i\int\mathcal{H}\,\mathcal{I}(y)\,dy}\right)\Phi_{p}\right) &= \left(\Phi_{p';\,\text{out}}^{+}V_{\alpha}(x)\,\Phi_{p;\,\text{in}}\right),\\ \left(\Phi_{p'}^{+}T\left(a_{\alpha}(x)\,e^{-i\int\mathcal{H}\,\mathcal{I}(y)\,dy}\right)\Phi_{p}\right) &= \left(\Phi_{p';\,\text{out}}^{+}A_{\alpha}(x)\,\Phi_{p;\,\text{in}}\right).\end{aligned}\right\}\tag{12.86}$$

Here $V_{\alpha}(x)$ and $A_{\alpha}(x)$ are operators of the vector and axial currents in the Heisenberg representation [see (12.29)] and $\Phi_{p;\,\text{in}}$ and $\Phi_{p;\,\text{out}}$ are the state vectors of the initial and final nucleons in this representation. Analogous to (12.32) we have

$$\left(\Phi_{p';\,\text{out}}^{+}\left(V_{\alpha}(x)+A_{\alpha}(x)\right)\Phi_{p;\,\text{in}}\right) = e^{-i(p'-p)\,x}\left(\Phi_{p';\,\text{out}}^{+}\left(V_{\alpha}(0)+A_{\alpha}(0)\right)\Phi_{p;\,\text{in}}\right).\tag{12.87}$$

Inserting (12.87) into (12.85) we obtain for the matrix element (the indices in and out henceforth will be omitted)

$$\begin{aligned}\left(\Phi_{q',\,p'}^{+}S^{G}\Phi_{q,\,p}\right) = (-i)\,&\frac{1}{(2\pi)^{3}}\left(\frac{m}{q_{0}'}\right)^{1/2}\frac{G}{\sqrt{2}}\,(2\pi)^{4}\,\delta(p'+q'-p-q)\\ &\times\left(\bar{u}(q')\,\gamma_{\alpha}(1+\gamma_{5})\,u(q)\right)\left(\Phi_{p'}^{+}\left(V_{\alpha}(0)+A_{\alpha}(0)\right)\Phi_{p}\right).\end{aligned}\tag{12.88}$$

If strong interactions are not considered, from (12.79) and (12.80) we find that the matrix element

$$\left(\Phi_{p'}^{+}V_{\alpha}(0)\,\Phi_{p}\right)\quad\text{and}\quad\left(\Phi_{p'}^{+}A_{\alpha}(0)\,\Phi_{p}\right)$$

are equal, respectively, to

$$\frac{1}{(2\pi)^{3}}\left(\frac{M^{2}}{p_{0}p_{0}'}\right)^{1/2}\bar{u}(p')\,\gamma_{\alpha}u(p)\quad\text{and}\quad\frac{1}{(2\pi^{3})}\left(\frac{M^{2}}{p_{0}p_{0}'}\right)^{1/2}\bar{u}(p')\,\gamma_{\alpha}\gamma_{5}u(p).$$

In the general case the matrix elements of the vector and axial currents can be written in the forms

$$\left.\begin{aligned}\left(\Phi_{p'}^{+}V_{\alpha}(0)\,\Phi_{p}\right) &= \frac{1}{(2\pi)^{3}}\left(\frac{M^{2}}{p_{0}p_{0}'}\right)^{1/2}\bar{u}(p)V_{\alpha}(p',p)\,u(p),\\ \left(\Phi_{p'}^{+}A_{\alpha}(0)\,\Phi_{p}\right) &= \frac{1}{(2\pi)^{3}}\left(\frac{M^{2}}{p_{0}p_{0}'}\right)^{1/2}\bar{u}(p')A_{\alpha}(p',p)\,u(p),\end{aligned}\right\}\tag{12.89}$$

where the matrix elements

$$\bar{u}(p')V_{\alpha}(p',p)\,u(p)\quad\text{and}\quad\bar{u}(p')A_{\alpha}(p',p)\,u(p)$$

transform respectively like a vector and a pseudovector.

By the same kind of argument as we used in the case of electron–neutron scattering, we conclude that the matrices $V_{\alpha}(p',p)$ and $A_{\alpha}(p',p)$ have the following general forms:

$$\left.\begin{aligned}V_{\alpha}(p',p) &= \gamma_{\alpha}G_{V}(\varkappa^{2})+iP_{\alpha}\frac{F_{V}(\varkappa^{2})}{2M}+i\varkappa_{\alpha}\frac{H_{V}(\varkappa^{2})}{2M},\\ A_{\alpha}(p',p) &= \gamma_{\alpha}\gamma_{5}G_{A}(\varkappa^{2})+i\varkappa_{\alpha}\gamma_{5}\frac{F_{A}(\varkappa^{2})}{2M}+iP_{\alpha}\gamma_{5}\frac{H_{A}(\varkappa^{2})}{2M}.\end{aligned}\right\}\tag{12.90}$$

Here $G_{V}, F_{V}, G_{A}, F_{A}, \ldots$ are quadratic functions of the four-momentum transfer $\varkappa^{2}$ ($\varkappa = p'-p$; $P = p'+p$).

It follows from the unitarity of the S-matrix and invariance under time-reversal that these functions are real. Actually, from the condition of unitarity of the S-matrix we obtain [see (12.50)]

$$(\Phi^+_{q',p'}R\Phi_{q,p})+(\Phi^+_{q,p}R\Phi_{q',p'})^* = -\sum_n(\Phi^+_{q',p'}R^+\Phi_n)(\Phi^+_n R\Phi_{q,p}). \tag{12.91}$$

The right-hand side of this relationship receives a contribution only from the states into which μ^-, p and γ_μ, n can transform. It is not difficult to see that the largest contribution is made by the state describing the μ-meson and the proton. In this case the matrix element $(\Phi^+_{q',p'}R^+\Phi_n)$ is of the order α (μ^--meson scattering on the proton) and $(\Phi^+_n R\Phi_{q,p})$ is of the order G [the process (12.78a)]. Omitting the quantities of order G_α in comparison with the quantities of order G we obtain from (12.91)

$$(\Phi^+_{q',p'}R\Phi_{q,p}) = -(\Phi^+_{q,p}R\Phi_{q',p'})^*. \tag{12.92}$$

This relationship relates the matrix elements of the direct and reverse processes.

Let us assume that there is invariance under time-reversal (T-invariance). T-invariance, as we know, implies that

$$(\Phi^+_{q,p}R\Phi_{q',p'}) = (\Phi^+_{q'_T,p'_T}R\Phi_{q_T,p_T}). \tag{12.93}$$

Here Φ_{q_T,p_T} is a state vector describing a neutron with momentum $p_T=(-\boldsymbol{p}, ip_0)$ and a neutrino with momentum $q_T=(-\boldsymbol{q}, iq_0)$. The spin projections of the neutron (neutrino) in the states $\Phi_{q,p}$ and Φ_{q_T,p_T} are opposite (under time-reversal the directions of the momenta and spins reverse).

In order to characterize the spin states of the particles described by the vectors Φ_{q_T,p_T} and $\Phi_{q'_T,p'_T}$, we examine the equations for the spinors $u(p)$ and $\bar{u}(p)$. We have

$$(\gamma p-iM)u(p) = 0, \tag{12.94a}$$

$$\bar{u}(p)(\gamma p-iM) = 0. \tag{12.94b}$$

From the second equation using the Hermiticity of the matrices γ_μ we find

$$(\gamma^* p-iM)\bar{u}(p) = 0. \tag{12.95}$$

We multiply this equation by the unitary matrix V_I which satisfies the condition

$$V_T\gamma^*_\mu V_T^{-1} = \gamma_\mu\delta_\mu \tag{12.96}$$

($\delta_\mu = -1$ for $\mu = 1, 2, 3$ and $\delta_4 = 1$; there is no summation over the index μ on the right-hand side of the relationship). We obtain

$$(\gamma p_T-iM)\, V_T\bar{u}(p) = 0, \tag{12.97}$$

where

$$p_T = (-\boldsymbol{p}, ip_0).$$

Assume

$$u_T(p_T) = V_T\bar{u}(p). \tag{12.98}$$

From (12.97) and (12.98) it follows that the spinor $u_T(p_T)$ satisfies the Dirac equation:

$$(\gamma p_T-iM)\, u_T(p_T) = 0.$$

With the help of (12.96) it is easy to verify that

$$\bar{u}_T(p_T) = u(p)\, V_T^{-1}. \tag{12.99}$$

We calculate the mean value of the spin operator

$$\Sigma_i = \tfrac{1}{2}(-i)\, \varepsilon_{ikl}\alpha_k\alpha_l = \tfrac{1}{2}(-i)\, \varepsilon_{ikl}\gamma_k\gamma_l$$

in the state $u_T(p_T)$. With the help of (12.96) we find

$$V_T\Sigma_i^* V_T^{-1} = -\Sigma_i. \tag{12.100}$$

Using (12.98)–(12.100) we obtain

$$\frac{(u_T^+(p_T)\, \Sigma_i u_T(p_T))}{(u_T^+(p_T)\, u_T(p_T))} = -\frac{(u^+(p)\, \Sigma_i u(p))}{(u^+(p)\, u(p))}, \tag{12.101}$$

i.e. the mean values of the spin operator Σ_i in the states $u(p)$ and $u_T(p_T)$ are equal in size and opposite in sign. Obviously all of this also applies to the spinor $u_T(q_T) = V_T\bar{u}(q)$.

We come to the conclusion that the spin states of the nucleon and neutrino described by the vector Φ_{p_T, q_T} are characterized by the spinors $u_T(p_T)$ and $u_T(q_T)$. Analogously the proton and μ-meson in the state $\Phi_{q_T', p_T'}$ have momenta $p_T' = (-\boldsymbol{p}', ip_0')$ and $q_T' = (-\boldsymbol{q}', iq_0')$ and are described by the spinors

$$u_T(p_T') = V_T\bar{u}(p') \quad \text{and} \quad u_T(q_T') = V_T\bar{u}(q').$$

From (12.92) and (12.93) we obtain

$$(\Phi_{q', p'}^+ R\Phi_{q, p}) = -(\Phi_{q_T', p_T'}^+ R\Phi_{q_T, p_T})^*. \tag{12.102}$$

Inserting (12.88) into (12.102) we obtain

$$\begin{aligned}(\bar{u}(q')\, \gamma_\alpha(1+\gamma_5)\, u(q))(\bar{u}(p')[V_\alpha(p', p)+A_\alpha(p', p)]\, u(p))\\ = (\bar{u}_T(q_T')\, \gamma_\alpha(1+\gamma_5)\, u_T(q_T))^*\, (\bar{u}_T(p_T')[V_\alpha(p_T', p_T)+A_\alpha(p_T', p_T)]\, u_T(p_T))^*.\end{aligned} \tag{12.103}$$

With the help of (12.96), (12.98), and (12.99) we find

$$(\bar{u}_T(q_T')\, \gamma_\alpha(1+\gamma_5)\, u_T(q_T))^* = (u(q')\, \gamma_\alpha^*(1-\gamma_5^*)\, \bar{u}(q))^*\, \delta_\alpha = (\bar{u}(q')\, \gamma_\alpha(1+\gamma_5)\, u(q)). \tag{12.104}$$

In obtaining (12.104) we used the relationship

$$V_T^{-1}\gamma_5 V_T = -\gamma_5^*,$$

whose validity is easy to verify from (12.96). Furthermore, from (12.90), (12.96), (12.98), and (12.99) we find

$$\left.\begin{aligned}(\bar{u}_T(p_T')\, V_\alpha(p_T', p_T)\, u_T(p_T))^* &= \bar{u}(p')\left[\gamma_\alpha G_V^* + iP_\alpha \frac{F_V^*}{2M} + i\varkappa_\alpha \frac{H_V^*}{2M}\right] u(p),\\ (\bar{u}_T(p_T')\, A_\alpha(p_T', p_T)\, u_T(p_T))^* &= \bar{u}(p')\left[\gamma_\alpha\gamma_5 G_A^* + i\varkappa_\alpha\gamma_5 \frac{F_A^*}{2M} + iP_\alpha\gamma_5 \frac{H_A^*}{2M}\right] u(p).\end{aligned}\right\} \tag{12.105}$$

With the help of (12.103), (12.104), and (12.105) we obtain

$$\left.\begin{aligned} G_V^* &= G_V, & F_V^* &= F_V, & H_V^* &= H_V, \\ G_A^* &= G_A, & F_A^* &= F_A, & H_A^* &= H_A. \end{aligned}\right\} \tag{12.106}$$

Thus, if there is T-invariance, all form factors in expressions (12.90) are real. Furthermore, using the isotopic invariance of strong interactions, we shall show that the form factors H_V and H_A vanish. Denote the sum of the vector and axial currents by $J_\alpha(x)$:

$$J_\alpha(x) = V_\alpha(x) + A_\alpha(x).$$

We shall call this operator the current operator.

Consider the operator

$$j_\alpha^0(x) = (\bar{\psi}_p(x)\,\gamma_\alpha(1+\gamma_5)\,\psi_n(x)). \tag{12.107}$$

Under the hypothesis of isotopic invariance the proton and neutron are two states of one particle—the nucleon. We denote the operator of the nucleon field by $\psi_\xi(x)$. We assume that $\psi_1(x) = \psi_p(x)$ and $\psi_{-1}(x) = \psi_n(x)$. The operator (12.107) can be written

$$j_\alpha^0 = (\bar{\psi}_p\gamma_\alpha(1+\gamma_5)\,\psi_n) = (\bar{\psi}_\xi\gamma_\alpha(1+\gamma_5)(\tau_+)_{\xi\xi'}\psi_{\xi'}) = (\bar{\psi}\gamma_\alpha(1+\gamma_5)\,\tau_+\psi). \tag{12.108}$$

Here

$$\tau_+ = \begin{pmatrix} 0 & 1 \\ 0 & 0 \end{pmatrix} = \tfrac{1}{2}(\tau_1 + i\tau_2), \tag{12.109}$$

where τ_1 and τ_2 are Pauli matrices.

The nucleon field operator $\psi_\xi^{(+)}(x)$ can be written in the form

$$\psi_\xi^{(+)}(x) = \frac{1}{(2\pi)^{3/2}} \int \left(\frac{M}{p_0}\right)^{1/2} u^r(p)\,\chi_\xi^\lambda c_{r,\lambda}(p)\,e^{ipx}\,d\mathbf{p}. \tag{12.110}$$

Here the index λ assumes the values p and n; $c_{r,\,p}(p)$ and $c_{r,\,n}(p)$ are respectively the annihilation operators of the proton and neutron and

$$\chi^p = \begin{pmatrix} 1 \\ 0 \end{pmatrix}, \quad \chi^n = \begin{pmatrix} 0 \\ 1 \end{pmatrix}. \tag{12.111}$$

Let us examine the transformation

$$\psi'(x) = U\psi(x)\,U^{-1} = e^{(i/2)\,\tau_2\pi}\psi(x) = i\tau_2\psi(x) \tag{12.112}$$

(rotation by the angle π around the 2 axis in isotopic space; U is a unitary operator acting on the field operators). From (12.110) and (12.112) we find (the spin index is omitted)

$$Uc_\lambda(p)\,U^{-1} = \sum_{\lambda'} ((\chi^\lambda)^+ i\tau_2\chi^{\lambda'})\,c_{\lambda'}(p). \tag{12.113}$$

Furthermore, with the help of (12.111) we obtain

$$Uc_p(p)\,U^{-1} = c_n(p), \quad Uc_n(p)\,U^{-1} = -c_p(p). \tag{12.114}$$

From this by Hermitian conjugation we find

$$Uc_p^+(p)\,U^{-1} = c_n^+(p); \quad Uc_n^+(p)\,U^{-1} = -c_p^+(p). \tag{12.115}$$

Noting that $U\Phi_0 = \Phi_0$ we obtain from (12.115)

$$Uc_p^+(p)\,\Phi_0 = c_n^+(p)\,\Phi_0, \quad Uc_n^+(p)\,\Phi_0 = -c_p^+(p)\,\Phi_0. \tag{12.116}$$

Thus, if the state vector describing a proton with momentum p is acted upon by the operator U, we obtain the vector describing a neutron with the same momentum. If the vector describing a neutron is acted upon by the operator U, we obtain (with a minus sign) the vector describing a proton. From (12.108) and (12.112) we find

$$\begin{aligned} Uj_\alpha^0(x)\,U^{-1} &= \bar{\psi}(x)\,\gamma_\alpha(1+\gamma_5)\,\tau_2\,\tfrac{1}{2}\,(\tau_1+i\tau_2)\,\tau_2\psi(x) \\ &= -\bar{\psi}(x)\,\gamma_\alpha(1+\gamma_5)\,\tfrac{1}{2}\,(\tau_1-i\tau_2)\,\psi(x) = -\left(j_\alpha^0(x)\right)^+\,\delta_\alpha. \end{aligned} \tag{12.117}$$

There is no summation over α on the right-hand side.

Assume that the total current $j_\alpha(x) = v_\alpha(x)+a_\alpha(x)$ in the weak interaction Hamiltonian (12.79) satisfies the relationship†

$$Uj_\alpha(x)U^{-1} = -j_\alpha^+(x)\,\delta_\alpha \tag{12.118}$$

[for the operator $\phi(x)$ corresponding to particles with isotopic spin I, $U\phi(x)U^{-1} = e^{i\pi I_2}\phi(x)$].

For the matrix element we are interested in we obtain

$$\begin{aligned} \left(\Phi_{p'(p)}^+ T\left(j_\alpha(0)\,\mathrm{e}^{-i\int \mathcal{H}_I^q(x)\,dx}\right)\Phi_{p(n)}\right) &= \left(\Phi_{p'(p)}^+ U^{-1}UT\left(j_\alpha(0)\,\mathrm{e}^{-i\int \mathcal{H}_I^q(x)\,dx}\right)U^{-1}U\Phi_{p(n)}\right) \\ &= \left(\Phi_{p'(n)}^+ T\left(j_\alpha^+(0)\,\delta_\alpha\,\mathrm{e}^{-i\int \mathcal{H}_I^q(x)\,dx}\right)\Phi_{p(p)}\right). \end{aligned} \tag{12.119}$$

Here $\Phi_{p'(p)}$ is the state vector describing a proton with momentum p', $\Phi_{p(n)}$ is the state vector of a neutron with momentum p, and so forth. In obtaining (12.119) we used the relationship

$$U\mathcal{H}_I^q U^{-1} = \mathcal{H}_I^q, \tag{12.120}$$

which follows from the isotopic invariance of the strong interaction Hamiltonian. Transforming to the Heisenberg representation we obtain from (12.119)

$$\left(\Phi_{p'(p)}^+ J_\alpha(0)\,\Phi_{p(n)}\right) = \left(\Phi_{p'(n)} J_\alpha^+(0)\,\Phi_{p(p)}\right)\delta_\alpha = \left(\Phi_{p(p)}^+ J_\alpha(0)\,\Phi_{p'(n)}\right)^*\delta_\alpha. \tag{12.121}$$

Inserting expression (12.89) into this relationship we find

$$\begin{aligned} \left(\bar{u}(p')\left(V_\alpha(p',p)+A_\alpha(p',p)\right)u(p)\right) &= \left(\bar{u}(p)\left(V_\alpha(p,p')+A_\alpha(p,p')\right)u(p')\right)^*\delta_\alpha \\ &= \left(\bar{u}(p')\left(\bar{V}_\alpha(p,p')+\bar{A}_\alpha(p,p')\right)u(p)\right)\delta_\alpha. \end{aligned} \tag{12.122}$$

† In the general case the current j_α can be written in the form

$$j_\alpha = j_\alpha^{\mathrm{I}}+j_\alpha^{\mathrm{II}},$$

where

$$j_\alpha^{\mathrm{I}} = \tfrac{1}{2}\,(j_\alpha - Uj_\alpha^+\,\delta_\alpha U^{-1}), \quad j_\alpha^{\mathrm{II}} = \tfrac{1}{2}\,(j_\alpha + Uj_\alpha^+\,\delta_\alpha U^{-1}).$$

It is obvious that

$$Uj_\alpha^{\mathrm{I}}U^{-1} = -\delta_\alpha(j_\alpha^{\mathrm{I}})^+, \quad Uj_\alpha^{\mathrm{II}}U^{-1} = \delta_\alpha(j_\alpha^{\mathrm{II}})^+.$$

The assumption (12.118) means that the current j_α^{II}—the so-called second-class current—does not enter the weak interaction Hamiltonian.

Furthermore, using (12.90) and (12.106) we obtain from (12.122)

$$H_V = 0 \quad \text{and} \quad H_A = 0. \tag{12.123}$$

Thus, if our assumptions are correct (T-invariance, non-existence of second-class currents), then the matrix element $(\Phi^+_{p'} J_\alpha(0) \Phi_p)$ is characterized by four real form factors.

It is not difficult to see that the fact that the form factor H_V vanishes is also a consequence of the hypothesis of the conservation of vector current. Actually, from the condition of the conservation of vector current

$$\frac{\partial V_\alpha}{\partial x_\alpha} = 0 \tag{12.124}$$

we obtain for the matrix element

$$(p'-p)_\alpha \left(\Phi^+_{p'} V_\alpha(0)\, \Phi_p\right) = 0. \tag{12.125}$$

Hence, using (12.89) and (12.90) we find

$$H_V = 0.$$

The diagram depicted in Fig. 19 corresponds to the matrix element of the process (12.78a) under discussion. The external lines of this diagram correspond to the spinors with the cor-

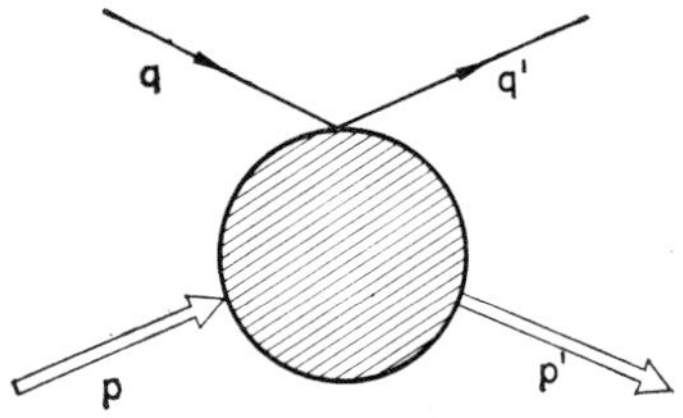

FIG. 19

responding normalized factors, the lepton vertex to the matrix $(G/\sqrt{2})^{1/2}\gamma_\alpha(1+\gamma_5)$, and the hatched part of the diagram, including strong interactions, corresponds to the matrix

$$\left(\frac{G}{\sqrt{2}}\right)^{1/2} \left[\gamma_\alpha(G_V+\gamma_5 G_A) + iP_\alpha \frac{F_V}{2M} + i\varkappa_\alpha \gamma_5 \frac{F_A}{2M}\right].$$

To conclude we shall obtain the matrix element of the process (12.78b). It is obvious that the second term of the Hamiltonian (12.79) contributes to the matrix element for this process. Denote by q and q' the four-momenta of the antineutrino and the μ^+-meson and by p and p' the four-momenta of the proton and neutron. In the first order in G the matrix element of the process is

$$(\Phi^+_{q',p'} S^G \Phi_{q,p}) = i\,\frac{2}{(2\pi)^3}\left(\frac{m}{q'_0}\right)^{1/2} (\bar{u}(-q)\,\gamma_\alpha(1+\gamma_5)\,u(-q))$$

$$\times (2\pi)^4\, \delta(p'+q'-p-q)\left(\Phi^+_{p'(n)} T\left(j^+_\alpha(0)\, \delta_\alpha e^{-i\int \mathscr{H}^{q}_{I}(x)\,dx}\right)\Phi_{p(p)}\right). \tag{12.126}$$

Using relationship (12.119) we find that

$$\left(\Phi^{+}_{p'(n)} T\left(j^{+}_{\alpha}(0)\,\delta_{\alpha} e^{-i\int \mathscr{H}^{a}_{I}(x)\,dx}\right)\Phi_{p(p)}\right)$$
$$= \frac{1}{(2\pi)^3}\left(\frac{M^2}{p_0 p'_0}\right)^{1/2}\left(\bar{u}(p')\left[\gamma_\alpha G_V + iP_\alpha \frac{F_V}{2M} + \gamma_\alpha\gamma_5 G_A + i\varkappa_\alpha\gamma_5 \frac{F_A}{2M}\right]u(p)\right). \quad (12.127)$$

Thus, the matrix elements of the processes (12.78a) and (12.78b) are characterized by the same form factors.

CHAPTER 6

CALCULATION OF THE CROSS SECTIONS AND DECAY PROBABILITIES

13. Cross Sections. Traces

In this section we shall define the cross section. We shall examine a specific process of the transformation of two particles into two particles. Denote the four-momenta of the initial (final) particles by p_1 and p_2 (p_1' and p_2'). The transition matrix element from the initial state Φ_i into the final state Φ_f we write in the form

$$(\Phi_f^+ S\Phi_i) = (\Phi_f^+\Phi_i) + (\Phi_f^+(S-1)\Phi_i). \tag{13.1}$$

The second term on the right-hand side of (13.1) is determined by the interaction; it vanishes if the interaction is absent and can always be written

$$(\Phi_f^+(S-1)\,\Phi_i) = \delta(P_f - P_i)\,R_{fi}. \tag{13.2}$$

Here $P_f = p_1' + p_2'$ is the total four-momentum of the final particles and $P_i = p_1 + p_2$ is the total four-momentum of the initial particles.

We are interested in the probability of the transition from the initial state into the final due to interaction. This probability is given by the square of the modulus of the second term in expression (13.1). Let dW_{fi} denote the probability of the final particles having momenta in the intervals from $\boldsymbol{p}_1'$ to $\boldsymbol{p}_1' + d\boldsymbol{p}_1'$ and from $\boldsymbol{p}_2'$ to $\boldsymbol{p}_2' + d\boldsymbol{p}_2'$. Then

$$dW_{fi} = |R_{fi}|^2\,\delta(P_f - P_i) \lim_{\substack{V\to\infty \\ T\to\infty}} \left(\frac{1}{(2\pi)^4} \int_{-T/2}^{T/2} dx_0 \int_V d\boldsymbol{x}\, e^{i(P_f - P_i)x} \right) d\boldsymbol{p}_1'\, d\boldsymbol{p}_2'. \tag{13.3}$$

The second δ-function is written in the form

$$\delta(P_f - P_i) = \frac{1}{(2\pi)^4} \lim_{\substack{T\to\infty \\ V\to\infty}} \int_{-T/2}^{T/2} dx_0 \int_V d\boldsymbol{x}\, e^{i(P_f - P_i)x}.$$

Since the integral in expression (13.3) is multiplied by $\delta(P_f - P_i)$, then $e^{i(P_f - P_i)x}$ under the integral sign can be replaced by 1. Then

$$dW_{fi} = |R_{fi}|^2\,\delta(P_f - P_i)\,d\boldsymbol{p}_1'\,d\boldsymbol{p}_2' \frac{1}{(2\pi)^4} \lim_{\substack{V\to\infty \\ T\to\infty}} VT. \tag{13.4}$$

Thus, $dW_{fi} \to \infty$. This is related to the fact that dW_{fi} is the transition probability during the time $-\infty$ to ∞ (the definition of the S-matrix) and in all space (the initial and final wave functions are plane waves). If V and T are sufficiently large, the transition probability for time T in volume V is

$$(dW_{fi})_{T,V} = |R_{fi}|^2\, \delta(P_f - P_i)\, d\boldsymbol{p}_1'\, d\boldsymbol{p}_2'\, \frac{1}{(2\pi)^4}\, VT. \tag{13.5}$$

From this the transition probability per unit time and unit volume under the condition that the transitions proceed in a sufficiently large volume and during a sufficiently long period of time is equal to

$$dW_{fi} = \frac{1}{(2\pi)^4}\, |R_{fi}|^2\, \delta(P_f - P_i)\, d\boldsymbol{p}_1'\, d\boldsymbol{p}_2'. \tag{13.6}$$

This quantity is finite and does not depend on V or T.

We now define the differential cross section of the process. We shall examine the volume element (Fig. 20) in the reference system where particles with momentum p_2 (target particles) are at rest (the laboratory system). Let the area of the hatched space perpendicular to the momentum of the incident particle in the laboratory system, $\boldsymbol{p}_1^L$, be equal to 1.

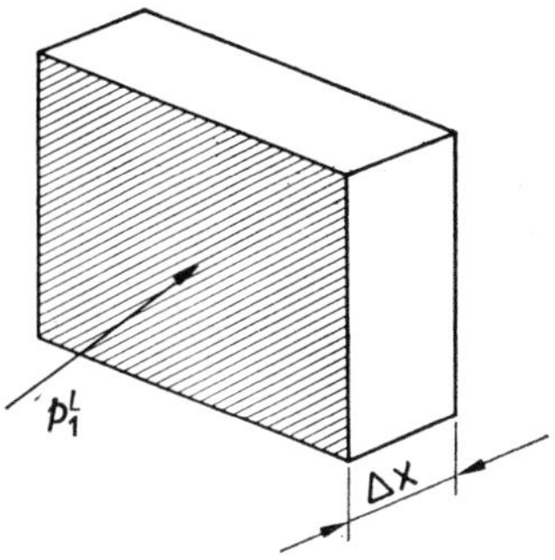

FIG. 20

The detection probability of target particles in this element of volume is equal to $\varrho_2^L \Delta x \cdot 1$, where ϱ_2^L is the probability density. We multiply this quantity by the probability that an incident particle will pass through the hatched unit area during a unit of time. We obtain

$$v_1^L \varrho_1^L \varrho_2^L \Delta x, \tag{13.7}$$

where v_1^L is the speed of the incident particles in the laboratory system and ϱ_1^L is the corresponding probability density. We call the quantity (13.7) the *flux* per unit volume and denote it by j. We have

$$j = \varrho_1^L \varrho_2^L v_1^L. \tag{13.8}$$

We write this quantity in an arbitrary reference system. It is not difficult to see that the speed of the incident particle in the laboratory system can be written

$$v_1^L = \frac{|\boldsymbol{p}_1^L|}{p_{10}^L} = \frac{\sqrt{(p_1 p_2)^2 - m_1^2 m_2^2}}{-(p_1 p_2)}. \tag{13.9}$$

Here p_1 and p_2 are the four-momenta of the incident particle and the target particle in an arbitrary reference system and m_1 and m_2 are the masses of the particles. Denote by $j_{1\mu}$ and $j_{2\mu}$ the four-vectors of flux $[j_k = \varrho_k v_k, (j_k)_4 = i\varrho_k$, where ϱ_k and v_k are, respectively, the density and speed, $k = 1, 2]$. We have

$$\varrho_1^L \varrho_2^L = -(j_1 j_2). \tag{13.10}$$

With the help of (13.8)–(13.10) we obtain

$$j = \frac{\sqrt{(p_1p_2)^2 - m_1^2 m_2^2}}{(p_1p_2)} (j_1 j_2). \tag{13.11}$$

Obviously this quantity is invariant. It is not difficult to see that

$$(j_1j_2) = \varrho_1\varrho_2 \left(\frac{\mathbf{p}_1\mathbf{p}_2}{p_{10}p_{20}} - 1\right) = \frac{\varrho_1\varrho_2(p_1p_2)}{p_{10}p_{20}}. \tag{13.12}$$

Ultimately we find

$$j = \frac{\sqrt{(p_1p_2)^2 - m_1^2 m_2^2}}{p_{10}p_{20}} \varrho_1\varrho_2. \tag{13.13}$$

The differential cross section is defined as the ratio of the transition probability per unit volume per unit time dW_{fi} to the flux j:

$$d\sigma_{fi} = \frac{dW_{fi}}{j}. \tag{13.14}$$

In so far as the four-dimensional volume is invariant, the probability dW_{fi} is also invariant. Thus, the cross section $d\sigma_{fi}$ defined by relationship (13.14) is a scalar.

We write the matrix element R_{fi} [see (13.2)] in the form

$$R_{fi} = \frac{1}{(2\pi)^6} \frac{1}{\sqrt{p_{10}p_{20}p'_{10}p'_{20}}} (2\pi)^4 M_{fi}. \tag{13.15}$$

We separate out the factors $(1/(2\pi)^{3/2})\,(1/\sqrt{p_0})$ related to each external line as well as the factor $(2\pi)^4$. The matrix element M_{fi} is a scalar (see the expressions for the matrix elements found in the preceding sections). In our normalization the densities ϱ_1 and ϱ_2 in (13.13) are

$$\varrho_1 = \varrho_2 = \frac{1}{(2\pi)^3}. \tag{13.16}$$

With the help of (13.13)–(13.16) we obtain for the differential cross section of the process

$$d\sigma_{fi} = \frac{1}{(2\pi)^2} \frac{1}{\sqrt{(p_1p_2)^2 - m_1^2 m_2^2}} |M_{fi}|^2 \, \delta(P_f - P_i) \frac{d\mathbf{p}'_1}{dp'_{10}} \frac{d\mathbf{p}'_2}{p'_{20}}. \tag{13.17}$$

All quantities on the right-hand side of (13.17) are scalars.† Note that

$$\sqrt{(p_1p_2)^2-m_1^2m_2^2} = p_{10}p_{20}\sqrt{(\boldsymbol{v}_1-\boldsymbol{v}_2)^2-(\boldsymbol{v}_1\times\boldsymbol{v}_2)^2}, \qquad (13.18)$$

where $\boldsymbol{v}_1 = \boldsymbol{p}_1/p_{10}$ and $\boldsymbol{v}_2 = \boldsymbol{p}_2/p_{20}$ are the speeds of the particles.

Up to this point we have assumed that the initial and final particles are found in definite spin states. In measuring the cross sections experimentally final particles are recorded with definite momenta but with all possible values of spin projection. Besides that, initial particles can be arbitrarily polarized and consequently their spin state must be described by density matrices.

As an example let us look at a process in which two particles with spin $\frac{1}{2}$ and two spin zero particles take part. We shall assume that the particles with four-momenta p_1 and p_1' have spin $\frac{1}{2}$ and the spins of the particles with momenta p_2 and p_2' are zero. The matrix element M_{fi} can be written in the form

$$M_{fi} = \bar{u}^{r'}(p_1')\,\mathfrak{M}(p_1', p_2'; p_1, p_2)\,u^r(p_1), \qquad (13.19)$$

where the spinors $\bar{u}^{r'}(p_1')$ and $u^r(p_1)$ correspond to the emerging and entering external lines and $\mathfrak{M}(p_1', p_2'; p_1, p_2)$ is a matrix acting on the spin variables. We shall find M_{fi}^*.

We have

$$M_{fi}^* = (\bar{u}^{r'}(p_1')\,\mathfrak{M}u^r(p_1))^* = \big((u^r(p_1))^+\,\mathfrak{M}^+\gamma_4 u^{r'}(p_1')\big) = (\bar{u}^r(p_1)\,\overline{\mathfrak{M}}u^{r'}(p_1')), \qquad (13.20)$$

where

$$\overline{\mathfrak{M}} = \gamma_4\mathfrak{M}^+\gamma_4. \qquad (13.21)$$

If the spin state of the initial particles is described by the density matrices‡

$$\varrho_{\sigma\sigma'}(p_1) = \sum_r u_\sigma^r(p_1)\,\bar{u}_{\sigma'}^r(p_1)\,\alpha_r \qquad (13.22)$$

(the quantities α_r give the spin state in the basis $u^r(p_1)$), then the cross section of the process under discussion is equal to

$$d\sigma_{fi} = \frac{1}{(2\pi)^2}\frac{1}{\sqrt{(p_1p_2)^2-m_1^2m_2^2}}\left[\sum_{r,r'}(\bar{u}^{r'}(p_1')\,\mathfrak{M}u^r(p_1))(\bar{u}^r(p_1)\,\overline{\mathfrak{M}}u^{r'}(p_1'))\,\alpha_r\right]$$
$$\times\,\delta(P_f-P_i)\frac{d\boldsymbol{p}_1'}{p_{10}'}\frac{d\boldsymbol{p}_2'}{p_{20}'}. \qquad (13.23)$$

† It is not difficult to see that $d\boldsymbol{p}/p_0$ is a scalar. Actually, let us examine the Lorentz transformation

$$p_x' = \frac{p_x-\beta p_0}{\sqrt{1-\beta^2}}, \quad p_y' = p_y, \quad p_z' = p_z, \quad p_0' = \frac{p_0-\beta p_x}{\sqrt{1-\beta^2}}.$$

Recognizing that $p_0 = \sqrt{m^2+\boldsymbol{p}^2}$ we obtain

$$dp_x' = \frac{dp_x\left(1-\beta\dfrac{p_x}{p_0}\right)}{\sqrt{1-\beta^2}} = \frac{dp_xp_0'}{p_0}, \quad dp_y' = dp_y, \quad dp_z' = dp_z,$$

i.e.

$$\frac{d\boldsymbol{p}'}{p_0'} = \frac{d\boldsymbol{p}}{p_0}.$$

‡ The spin density matrix for the particles with spin $\frac{1}{2}$ is examined in the Appendix.

Using the relationship

$$\sum_r u^r_\sigma(p_1')\,\bar{u}^r_{\sigma'}(p_1') = \Lambda_{\sigma\sigma'}(p_1') = \left(\frac{\hat{p}_1'+im_1'}{2im_1'}\right)_{\sigma\sigma'} \tag{13.24}$$

(see Appendix) we find that the quantity in square brackets in expression (13.23) is equal to

$$\sum_{\sigma,\,\sigma',\,\sigma'',\,\sigma'''} \mathfrak{M}_{\sigma\sigma'}\left(\sum_r u^r_{\sigma'}(p_1)\,\bar{u}^r_{\sigma''}(p_1)\,\alpha_r\right)\overline{\mathfrak{M}}_{\sigma''\sigma'''}\left(\sum_{r'} u^{r'}_{\sigma'''}(p_1')\,\bar{u}^{r'}_{\sigma}(p_1')\right)$$

$$= \sum_\sigma (\mathfrak{M}\varrho(p_1)\,\overline{\mathfrak{M}}\Lambda(p_1'))_{\sigma\sigma} = \mathrm{Tr}\,[\mathfrak{M}\varrho(p_1)\,\overline{\mathfrak{M}}\Lambda(p_1')]. \tag{13.25}$$

Ultimately we find for the cross section of the process with polarized initial particles the following expression:

$$d\sigma_{fi} = \frac{1}{(2\pi)^2}\frac{1}{\sqrt{(p_1p_2)^2-m_1^2m_2^2}}\,\mathrm{Tr}\,[\mathfrak{M}\varrho(p_1)\,\overline{\mathfrak{M}}\Lambda(p_1')]\,\delta(P_f-P_i)\frac{dp_1'}{p_{10}'}\frac{dp_2'}{p_{20}'}. \tag{13.26}$$

To calculate the cross sections it is necessary therefore to know how to calculate the traces of the products of the γ-matrices. The methods for calculating the traces of products of γ-matrices are based on the relationships

$$\left.\begin{aligned} \gamma_\mu\gamma_\nu+\gamma_\nu\gamma_\mu &= 2\delta_{\mu\nu}, \\ \gamma_\mu\gamma_5+\gamma_5\gamma_\mu &= 0, \qquad \gamma_5^2 = 1, \end{aligned}\right\} \tag{13.27}$$

as well as the relationship

$$\mathrm{Tr}\,(AB) = \mathrm{Tr}\,(BA), \tag{13.28}$$

whose validity for any matrices A and B is not difficult to verify. Actually,

$$\mathrm{Tr}\,(AB) = \sum_{\sigma,\,\sigma'} A_{\sigma\sigma'}B_{\sigma'\sigma} = \sum_{\sigma,\,\sigma'} B_{\sigma'\sigma}A_{\sigma\sigma'} = \mathrm{Tr}\,(BA).$$

With the help of (13.27) and (13.28) we shall prove the following assertions:

1. The trace of the product of an odd number of γ-matrices is zero. Actually,

$$\mathrm{Tr}\,(\underbrace{\gamma_\mu\gamma_\nu\ldots\gamma_\varrho}_{\text{odd number}}) = \mathrm{Tr}\,(\gamma_5\gamma_5\gamma_\mu\gamma_\nu\ldots\gamma_\varrho) = -\mathrm{Tr}\,(\gamma_5\gamma_\mu\gamma_\nu\ldots\gamma_\varrho\gamma_5) = -\mathrm{Tr}\,(\gamma_\mu\gamma_\nu\ldots\gamma_\varrho) = 0. \tag{13.29}$$

The first equation in (13.29) was obtained by replacing 1 by γ_5^2, the second by permuting the matrix γ_5 with the matrices γ_μ, γ_ν, and so forth. The third equation was obtained with the help of relationship (13.28).

2.
$$\mathrm{Tr}\,(\gamma_\mu\gamma_\nu) = 4\delta_{\mu\nu}. \tag{13.30}$$

Actually, using (13.27) and (13.28) we find

$$\mathrm{Tr}\,(\gamma_\mu\gamma_\nu) = \mathrm{Tr}\,[2\delta_{\mu\nu}-\gamma_\nu\gamma_\mu] = 8\delta_{\mu\nu}-\mathrm{Tr}\,\gamma_\mu\gamma_\nu.$$

Let a and b be arbitrary four-vectors. Multiplying (13.30) by a_μ and b_ν and summing over μ and ν we obtain

$$\mathrm{Tr}\,(\hat{a}\hat{b}) = 4\,(ab). \tag{13.31}$$

3. $$\mathrm{Tr}\,(\gamma_\mu\gamma_\nu\gamma_\varrho\gamma_\sigma) = 4(\delta_{\mu\nu}\delta_{\varrho\sigma} - \delta_{\mu\varrho}\delta_{\nu\sigma} + \delta_{\mu\sigma}\delta_{\nu\varrho}). \tag{13.32}$$

We have

$$\begin{aligned}\mathrm{Tr}\,(\gamma_\mu\gamma_\nu\gamma_\varrho\gamma_\sigma) &= \mathrm{Tr}\,[(2\delta_{\mu\nu} - \gamma_\nu\gamma_\mu)\,\gamma_\varrho\gamma_\sigma] = 8\delta_{\mu\nu}\delta_{\varrho\sigma} - \mathrm{Tr}\,[\gamma_\nu(2\delta_{\mu\varrho} - \gamma_\varrho\gamma_\mu)\,\gamma_\sigma] \\ &= 8\delta_{\mu\nu}\delta_{\varrho\sigma} - 8\delta_{\mu\varrho}\delta_{\nu\sigma} + \mathrm{Tr}\,[\gamma_\nu\gamma_\varrho(2\delta_{\mu\sigma} - \gamma_\sigma\gamma_\mu)] = 8\delta_{\mu\nu}\delta_{\varrho\sigma} - 8\delta_{\mu\varrho}\delta_{\nu\sigma} + 8\delta_{\mu\sigma}\delta_{\nu\varrho} - \mathrm{Tr}\,(\gamma_\mu\gamma_\nu\gamma_\varrho\gamma_\sigma).\end{aligned}$$

Transferring the last term to the left we obtain (13.32). The method for calculation consists in the following. With the help of commutation relations (13.27) the matrix γ_μ is "pulled through" all remaining matrices, then relationship (13.28) is used to return matrix γ_μ to the initial position. It is obvious that in this way we can calculate the trace of the products of six, eight, and so forth γ-matrices.

Let a, b, c, and d be arbitrary four-vectors. Multiplying (13.32) by a_μ, b_ν, c_ϱ, and d_σ and summing over μ, ν, ϱ, and σ we obtain

$$\mathrm{Tr}\,(\hat{a}\hat{b}\hat{c}\hat{d}) = 4[(ab)(cd) - (ac)(bd) + (ad)(bc)]. \tag{13.33}$$

4. $$\mathrm{Tr}\,\gamma_5 = 0. \tag{13.34}$$

Actually,

$$\mathrm{Tr}\,\gamma_5 = \mathrm{Tr}\,(\gamma_1\gamma_1\gamma_5) = -\mathrm{Tr}\,(\gamma_1\gamma_5\gamma_1) = -\mathrm{Tr}\,\gamma_5 = 0.$$

5. $$\mathrm{Tr}\,(\gamma_5\gamma_\mu\gamma_\nu) = 0. \tag{13.35}$$

We write $\mathrm{Tr}\,(\gamma_5\gamma_\mu\gamma_\nu)$ in the form

$$\mathrm{Tr}\,(\gamma_5\gamma_\mu\gamma_\nu) = \tfrac{1}{2}\,\mathrm{Tr}\,[\gamma_5(\gamma_\mu\gamma_\nu - \gamma_\nu\gamma_\mu)] + \tfrac{1}{2}\,\mathrm{Tr}\,[\gamma_5(\gamma_\mu\gamma_\nu + \gamma_\nu\gamma_\mu)]. \tag{13.36}$$

The second term as a result of (13.27) is equal to $\delta_{\mu\nu}\,\mathrm{Tr}\,\gamma_5 = 0$. It is not difficult to verify the validity of the relationship

$$\gamma_5(\gamma_\mu\gamma_\nu - \gamma_\nu\gamma_\mu) = -\,\varepsilon_{\mu\nu\varrho\sigma}\gamma_\varrho\gamma_\sigma. \tag{13.37}$$

With the help of (13.37) and (13.30) we obtain

$$\mathrm{Tr}\,(\gamma_5\gamma_\mu\gamma_\nu) = -\tfrac{1}{2}\varepsilon_{\mu\nu\varrho\sigma}\,\mathrm{Tr}\,(\gamma_\varrho\gamma_\sigma) = -2\varepsilon_{\mu\nu\varrho\sigma}\delta_{\varrho\sigma} = 0.$$

6. $$\mathrm{Tr}\,(\gamma_5\gamma_\mu\gamma_\nu\gamma_\varrho\gamma_\sigma) = 4\varepsilon_{\mu\nu\varrho\sigma}. \tag{13.38}$$

Actually, it is not difficult to see that the trace on the left-hand side of this equation is antisymmetric under the transposition of any two indices:

$$\mathrm{Tr}\,[\gamma_5\gamma_\mu\gamma_\nu\gamma_\varrho\gamma_\sigma] = \mathrm{Tr}\,[\gamma_5\gamma_\mu(2\delta_{\nu\varrho} - \gamma_\varrho\gamma_\nu)\,\gamma_\sigma] = -\mathrm{Tr}\,(\gamma_5\gamma_\mu\gamma_\varrho\gamma_\nu\gamma_\sigma)$$

and so forth. Furthermore,

$$\mathrm{Tr}\,(\gamma_5\gamma_1\gamma_2\gamma_3\gamma_4) = \mathrm{Tr}\,\gamma_5^2 = 4.$$

14. Electron Scattering on a Nucleon

The electromagnetic form factors of nucleons are determined from experiment by scattering electrons on nucleons. In this section the cross section for this process will be calculated. Denote by q and q' (p and p') the four-momenta of the initial and final electrons (nucleons). The diagram of the process in the second order of perturbation theory in e is shown in Fig. 18. All strong interactions of the nucleon (the hatched part of the diagram) are considered phenomenologically through the introduction of form factors.

The matrix element of the process of electron–nucleon scattering is given by the following expression (see § 12):

$$(\Phi^+_{q',p'}S^{e^2}\Phi_{q,p}) = \frac{1}{(2\pi)^6}\, i \left(\frac{m^2M^2}{p_0p_0'q_0q_0'}\right)^{1/2} \frac{1}{(2\pi)^4}$$

$$\times(2\pi)^8e^2(\bar{u}(q')\,\gamma_\mu u(q))\,(\bar{u}(p')\Gamma_\mu(p',p)\,u(p))\,\frac{1}{\varkappa^2}\,\delta(p'+q'-p-q). \tag{14.1}$$

In this expression m and M are respectively the masses of the electron and the nucleon; $\varkappa = p'-p$ and

$$\Gamma_\mu(p',p) = \gamma_\mu F_1(\varkappa^2) - \frac{1}{2M}\sigma_{\mu\nu}\varkappa_\nu F_2(\varkappa^2), \tag{14.2}$$

where F_1 and F_2 are the Dirac and Pauli form factors of the nucleon.

With the help of the Dirac equation for the spinors $u(p)$ and $\bar{u}(p')$ it is not difficult to see that

$$\bar{u}(p')\,\sigma_{\mu\nu}\varkappa_\nu u(p) = -\bar{u}(p')\,[i(p'+p)_\mu + 2M\gamma_\mu]\,u(p). \tag{14.3}$$

Using this relationship we obtain

$$\bar{u}(p')\,\Gamma_\mu(p',p)\,u(p) = \bar{u}(p')\left[\gamma_\mu(F_1+F_2) + \frac{i}{2M}F_2(p'+p)_\mu\right]u(p). \tag{14.4}$$

As was shown in the preceding section, the calculation of the cross section reduces to the calculation of traces. Before doing this it is natural to decrease, in so far as possible, the number of γ-matrices in the matrix element with the help of the Dirac equation which the spinors of the initial and final particles obey. We use relationship (14.4). From (14.1), (13.5), and (14.4) we obtain for $M_{q'p';\,qp}$:

$$M_{q'p';\,qp} = ie^2mM(\bar{u}(q')\,\gamma_\mu u(q))\,(\bar{u}(p')\,R_\mu u(p))\,\frac{1}{\varkappa^2}\,. \tag{14.5}$$

Here

$$R_\mu(p',p) = \gamma_\mu G_M + i\,\frac{F_2}{M}\,P_\mu, \tag{14.6}$$

where

$$P = p+p',$$

and

$$G_M = F_1+F_2 \tag{14.7}$$

is the magnetic form factor.

Let us examine the scattering of unpolarized particles. In this case the initial density matrices of the electron and nucleon are equal to

$$\varrho(q) = \tfrac{1}{2}\,\Lambda(q), \quad \varrho(p) = \tfrac{1}{2}\,\Lambda(p), \tag{14.8}$$

where

$$\Lambda(q) = \frac{\hat{q}+im}{2im}, \qquad \Lambda(p) = \frac{\hat{p}+iM}{2im}$$

are projection operators (see Appendix).

From (14.7), (13.19), and (13.26) we obtain the following expression for the differential cross section of the process:

$$d\sigma = \frac{1}{(2\pi)^2}\frac{1}{\sqrt{(pq)^2-M^2m^2}}\, e^4m^2M^2\tfrac{1}{4}\frac{1}{(\varkappa^2)^2}\,\mathrm{Tr}\,[\gamma_\mu\Lambda(q)\,\bar{\gamma}_\nu\Lambda(q')]$$
$$\times\mathrm{Tr}\,[R_\mu\Lambda(p)\,\bar{R}_\nu\Lambda(p')]\,\delta(p'+q'-p-q)\frac{dp'\,dq'}{p_0'\;q_0'}. \tag{14.9}$$

Here

$$\bar{\gamma}_\nu = \gamma_4\gamma_\nu\gamma_4 = \delta_\nu\gamma_\nu; \qquad \bar{R}_\nu = \gamma_4R_\nu^+\gamma_4 = \left(\gamma_\nu G_M+iP_\nu\frac{F_2}{2M}\right)\delta_\nu \tag{14.10}$$

[remember that $\delta_\nu = -1$ for $\nu = 1, 2, 3$, and $\delta_4 = 1$; there is no summation over ν on the right-hand side of (14.10)].

With the help of (14.10) we have

$$\mathrm{Tr}\,[\gamma_\mu\Lambda(q)\,\bar{\gamma}_\nu\Lambda(q')]\;\mathrm{Tr}\,[R_\mu\Lambda(p)\,\bar{R}_\nu\Lambda(p')] = \mathrm{Tr}\,[\gamma_\mu\,\Lambda(q)\,\gamma_\nu\Lambda(q')]\;\mathrm{Tr}\,[R_\mu\Lambda(p)\,R_\nu\Lambda(p')]. \tag{14.11}$$

We now turn to the calculation of the traces. Using (13.29), (13.30), and (13.22) it is easy to find

$$\mathrm{Tr}\,[R_\mu\Lambda(p)R_\nu\Lambda(p')] = -\frac{1}{M^2}\Bigg\{G_M^2[p_\mu p_\nu'+p_\mu'p_\nu-\delta_{\mu\nu}(pp'+M^2)]$$
$$+P_\mu P_\nu\left[-2MG_M\left(\frac{F_2}{2M}\right)-\left(\frac{F_2}{2M}\right)^2(pp'-M^2)\right]\Bigg\}. \tag{14.12}$$

We shall write expression (14.12) in another form. Express p' and p in terms of P and $\varkappa$. We obtain

$$p' = \tfrac{1}{2}\,(P+\varkappa), \quad p = \tfrac{1}{2}\,(P-\varkappa). \tag{14.13}$$

From this we find

$$p_\mu p_\nu'+p_\nu p_\mu' = \tfrac{1}{2}\,(P_\mu P_\nu-\varkappa_\mu\varkappa_\nu). \tag{14.14}$$

It is further obvious that

$$pp'+M^2 = -\tfrac{1}{2}\,\varkappa^2, \quad pp'-M^2 = -\tfrac{1}{2}\,(\varkappa^2+4M^2). \tag{14.15}$$

Let us express F_2 in magnetic and charge form factors. We have

$$G_M = F_1+F_2, \quad G_E = F_1-\frac{\varkappa^2}{4M^2}F_2,$$

from which

$$\frac{1}{2M}F_2 = \frac{2(G_M-G_E)\,M}{\varkappa^2+4M^2}. \tag{14.16}$$

Using (14.14)–(14.16) we obtain

$$\operatorname{Tr}[R_\mu \Lambda(p)\, R_\nu \Lambda(p')] = -\frac{1}{2M^2}\left\{ G_M^2[\delta_{\mu\nu}\varkappa^2 - \varkappa_\mu \varkappa_\nu] + \frac{P_\mu P_\nu}{\varkappa^2+4M^2}[G_M^2\varkappa^2 + 4M^2 G_E^2]\right\}. \quad (14.17)$$

Thus, if one introduces magnetic and charge form factors, the interference term $G_M G_E$ does not enter the expression for the cross section. The latter is easy to understand with the help of relationship (12.67) (the expression for the cross section of the scattering of unpolarized particles cannot contain an interference term between the coefficient of the unit matrix in the amplitude and the coefficient of σ_i). Furthermore, in expression (14.17) the magnetic form factor is multiplied by quantities quadratically dependent on $\varkappa$, which is also easy to understand with the help of (12.67).

The electron trace can be obtained from (14.12). To do so, in expression (14.12) we must set $F_2 = 0$, $G_M = 1$, and make the replacement $p \to q$, $p' \to q'$, $M \to m$. We obtain

$$\operatorname{Tr}[\gamma_\mu \Lambda(q)\, \gamma_\nu \Lambda(q')] = -\frac{1}{m^2}[q_\mu q'_\nu + q_\nu q'_\mu + \tfrac{1}{2}\delta_{\mu\nu}\varkappa^2]. \quad (14.18)$$

In obtaining (14.18) we also used the relationship

$$qq' + m^2 = -\tfrac{1}{2}\varkappa^2, \quad (14.19)$$

which follows from the equation

$$q - q' = p' - p = \varkappa. \quad (14.20)$$

The next step is to multiply (14.18) by (14.17) and sum over μ and ν. Before performing these contractions we shall show that

$$\operatorname{Tr}[\gamma_\mu \Lambda(q)\, \gamma_\nu \Lambda(q')\, \varkappa_\mu] = 0. \quad (14.21)$$

It is easy to see that

$$\hat{q}\Lambda(q) = \frac{1}{2im}(q^2 + im\hat{q}) = im\Lambda(q), \quad \Lambda(q')\,\hat{q}' = im\Lambda(q'), \quad (14.22)$$

from which

$$\operatorname{Tr}[\gamma_\mu \Lambda(q)\, \gamma_\nu \Lambda(q')\, \varkappa_\mu] = \operatorname{Tr}[\Lambda(q')\,\hat{\varkappa}\Lambda(q)\,\gamma_\nu] = \operatorname{Tr}[\Lambda(q')\,(\hat{q} - \hat{q}')\,\Lambda(q)\,\gamma_\nu] = 0.$$

In an analogous manner we find

$$\operatorname{Tr}[\gamma_\mu \Lambda(q)\, \gamma_\nu \Lambda(q')\, \varkappa_\nu] = 0. \quad (14.23)$$

From these relationships it follows that the term $\varkappa_\mu \varkappa_\nu$ in (14.17) does not contribute to the cross section. In addition, if the four-vector P is put into the form $P = 2p + \varkappa$, we conclude from (14.21) and (14.23) that in contracting the tensors (14.17) and (14.18), $P_\mu P_\nu$ can be replaced by $4p_\mu p_\nu$. As a result we obtain

$$\begin{aligned} A &= \operatorname{Tr}[\gamma_\mu \Lambda(q)\, \gamma_\nu \Lambda(q')] \operatorname{Tr}[R_\mu \Lambda(p)\, R_\nu \Lambda(p')] \\ &= \frac{1}{2M^2 m^2}\left[G_M^2 \varkappa^2(\varkappa^2 - 2m^2) + \frac{4(G_M^2\varkappa^2 + 4M^2 G_E^2)}{\varkappa^2 + 4M^2}(2(pq)(pq') - \tfrac{1}{2}M^2\varkappa^2)\right]. \quad (14.24)\end{aligned}$$

We shall calculate the cross section for electron–nucleon scattering in the reference system in which the initial nucleon is at rest (the laboratory system). We first perform integration over $\boldsymbol{p}'$ (using the law of momentum conservation). We obtain

$$d\sigma = \frac{e^4}{4(2\pi)^2} \frac{M^2m^2}{\sqrt{(pq)^2-M^2m^2}} A \frac{1}{(\varkappa^2)^2} \delta(p_0'+q_0'-p_0-q_0) \frac{q_0'|\boldsymbol{q}'|\,dq_0'\,d\Omega'}{p_0'q_0'}. \tag{14.25}$$

Here $\boldsymbol{p}' = \boldsymbol{q}-\boldsymbol{q}'$ and the quantity A is given by expression (14.24). The element of volume $d\boldsymbol{q}'$ is written in spherical coordinates

$$d\boldsymbol{q}' = |\boldsymbol{q}'|^2\,d|\boldsymbol{q}'|\,d\Omega' = |\boldsymbol{q}'|\,q_0'\,dq_0'\,d\Omega'. \tag{14.26}$$

We integrate over q_0' (using the law of energy conservation). Here it is necessary to keep in mind that q_0' is found in the argument of the δ-function also through $p_0'=\sqrt{M^2+(\boldsymbol{q}-\boldsymbol{q}')^2}$. We use the relationship

$$\delta(f(x)) = \frac{\delta(x-x_0)}{|f'(x_0)|}, \tag{14.27}$$

where x_0 is the root of the function $f(x)$. We shall calculate the derivative with respect to q_0' of the argument of the δ-function

$$\frac{d}{dq_0'}\left(\sqrt{M^2+(\boldsymbol{q}-\boldsymbol{q}')^2}+q_0'\right) = 1+\frac{1}{p_0'}\left(q_0'-\frac{|\boldsymbol{q}|}{|\boldsymbol{q}'|}\,q_0'\cos\theta\right), \tag{14.28}$$

where θ is the angle between the momenta of the incident and scattered electrons (between $\boldsymbol{q}$ and $\boldsymbol{q}'$).

Let us examine the scattering of high-energy electrons on nucleons. Assume that $q_0 \gg m$. Omitting the terms of order m^2/q_0^2 and $m^2/q_0'^2$ we obtain

$$\left.\begin{aligned} &\varkappa^2 = 4q_0q_0'\sin^2\frac{\theta}{2}, \qquad \sqrt{(pq^2)-M^2m^2} = Mq_0,\\ &2(pq)(pq')-\tfrac{1}{2}M^2\varkappa^2 = 2M^2q_0q_0'\cos^2\frac{\theta}{2},\\ &\frac{d}{dq_0'}(p_0'+q_0') = \frac{M}{p_0'}\left(1+\frac{2q_0}{M}\sin^2\frac{\theta}{2}\right) \end{aligned}\right\} \tag{14.29}$$

(the derivative is calculated for the value of q' which follows from the law of energy conservation). Using (14.29), we ultimately obtain from (14.24) and (14.25) the following expression for the differential cross section for the scattering of high-energy electrons by nucleons in the laboratory system (the Rosenbluth formula):

$$\frac{d\sigma}{d\Omega} = \left(\frac{d\sigma}{d\Omega}\right)_0 \left[\frac{G_E^2+\dfrac{\varkappa^2}{4M^2}G_M^2}{1+\dfrac{\varkappa^2}{4M^2}}+\frac{\varkappa^2}{2M^2}G_M^2\tan^2\frac{\theta}{2}\right], \tag{14.30}$$

where

$$\left(\frac{d\sigma}{d\Omega}\right)_0 = \frac{\alpha^2 \cos^2 \dfrac{\theta}{2}}{4q_0^2 \sin^4 \dfrac{\theta}{2}\left(1+\dfrac{2q_0}{M}\sin^2\dfrac{\theta}{2}\right)}, \quad \alpha = \frac{e^2}{4\pi} = \frac{1}{137}. \tag{14.31}$$

Comparing (14.30) with the electron–nucleon scattering data we can determine the magnetic and charge form factors of the nucleon for various values of $\varkappa^2$.

Note that for $q_0 \gg m$

$$\varkappa^2 = \frac{4q_0^2 \sin^2 \dfrac{\theta}{2}}{1+2\dfrac{q_0}{M}\sin^2\dfrac{\theta}{2}}.$$

It is easy to obtain this expression from the laws of energy and momentum conservation.

In conclusion we shall show that $(d\sigma/d\Omega)_0$ [formula (14.31)] is the cross section for electrons with $q_0 \gg m$ scattering on a spin zero point particle.

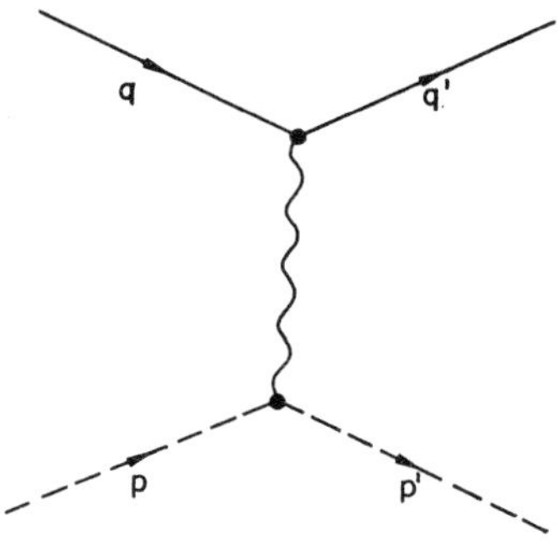

FIG. 21

The diagram of this process is shown in Fig. 21. The matrix element corresponding to the diagram in Fig. 21 can be obtained from (14.1) by replacing

$$\left(\frac{M^2}{p_0 p_0'}\right)^{1/2} \bar{u}(p')\,\Gamma_\mu(p', p)\,u(p)$$

by $(-i)/\sqrt{4p_0p_0'}$ [see expression (12.77); for a point particle the form factor is equal to one]. This means that the contraction A in (14.25) is equal to

$$\frac{2}{M^2m^2}\,[2(pq)(pq') - \tfrac{1}{2}M^2\varkappa^2].$$

In the laboratory system $q_0 \gg m$ we obtain expression (14.31) for the cross section.

15. The Process $\nu_\mu + n \to \mu^- + p$

We shall calculate the differential cross sections for the following processes:

$$\nu_\mu + n \to \mu^- + p, \tag{15.1a}$$
$$\bar{\nu}_\mu + p \to \mu^+ + n. \tag{15.1b}$$

These processes are of great interest in weak interaction physics and at the present time are being studied experimentally. We begin with an examination of the process (15.1a). Denote by q and q' the four-momenta of the initial neutrino and final μ-meson and by p and p' the four-momenta of the neutron and proton. The diagram of the process is shown in Fig. 19.

The effective weak interaction Hamiltonian was discussed in § 12. The matrix element of the process (15.1a) is equal to [see (12.88)–(12.90), (12.123)]

$$(\Phi^+_{q',\,p'} S \Phi_{q,\,p}) = (-i)\frac{1}{(2\pi)^6}\left(\frac{mM^2}{p_0 p_0' q_0'}\right)^{1/2}(2\pi)^4 \frac{G}{\sqrt{2}}\,\delta(p'+q'-p-q)$$
$$\times (\bar{u}(q')\gamma_\alpha(1+\gamma_5)u(q))(\bar{u}(p')[V_\alpha(p',p)+A_\alpha(p',p)]u(p)). \tag{15.2}$$

Here M and m are, respectively, the masses of the nucleon and the μ-meson; $G = 10^{-5}M^{-2}$ is the weak interaction constant and

$$\left.\begin{aligned} V_\alpha(p',p) &= \gamma_\alpha G_V(\varkappa^2) + iP_\alpha \frac{F_V(\varkappa^2)}{2M},\\ A_\alpha(p',p) &= \gamma_\alpha\gamma_5 G_A(\varkappa^2) + i\varkappa_\alpha\gamma_5 \frac{F_A(\varkappa^2)}{2M},\\ P = p+p', &\quad \varkappa = p'-p. \end{aligned}\right\} \tag{15.3}$$

The strong interactions of nucleons (the hatched part of the diagram in Fig. 19) are allowed for by the form factors G_V, F_V, G_A, and F_A. If the T-invariance is valid, all form factors are real (see § 12). The form factors G_V and F_V (G_A and F_A) characterize the matrix element of the vector part (the axial part) of the weak current. Note that the spinor $u(q)$ describing the neutrino is normalized with the condition $u^+(q)\,u(q) = 1$ (see Appendix).

From (15.2) we find that the matrix element $M_{q',\,p';\,q,\,p}$ defined by relationship (13.15) is equal to†

$$M_{q',\,p';\,q,\,p} = (-i)\sqrt{M^2 m q_0}\,\frac{G}{\sqrt{2}}\,(\bar{u}(q')\gamma_\alpha(1+\gamma_5)u(q))(\bar{u}(p')(V_\alpha + A_\alpha)u(p)). \tag{15.4}$$

† We observe that the contribution of the second term in the expression for $A_\alpha(p',p)$ to the matrix element of the process is

$$(\bar{u}(q')\gamma_\alpha(1+\gamma_5)u(q))(\bar{u}(p')i\gamma_5\varkappa_\alpha u(p))\frac{F_A}{2M} = (\bar{u}(q')(\hat{q}-\hat{q}')(1+\gamma_5)u(q))(\bar{u}(p')\gamma_5 u(p))\,i\,\frac{F_A}{2M}$$
$$= (\bar{u}(q')(1+\gamma_5)u(q))(\bar{u}(p')\gamma_5 u(p))\frac{mF_A}{2M}.$$

In obtaining this expression we used the law of four-momentum conservation ($\varkappa = p'-p = q-q'$) and the Dirac equation for the spinors $\bar{u}(q')$ and $u(q)$. Thus, the form factor $mF_A/2M$ characterizes the induced pseudoscalar interaction.

In accordance with present theory of weak interactions the neutrino is a particle with definite (negative) helicity. The density matrix of the neutrino is (see Appendix)

$$\varrho(q) = \tfrac{1}{2}(1+\gamma_5)\frac{\hat{q}}{2iq_0}.$$

We shall suppose that the initial neutrons are unpolarized, i.e. we assume that

$$\varrho(p) = \tfrac{1}{2}\Lambda(p) = \frac{\hat{p}+iM}{4iM}.$$

With the help of (13.26) we find the following expression for the differential cross section of the process:

$$d\sigma = \frac{1}{(2\pi)^2}\frac{1}{|(pq)|}M^2 mq_0\frac{G^2}{2}\frac{1}{2}\operatorname{Tr}\left[\gamma_\alpha(1+\gamma_5)\frac{\hat{q}}{2iq_0}\overline{\gamma_\beta(1+\gamma_5)}\Lambda(q')\right]$$
$$\times \operatorname{Tr}\left[(V_\alpha+A_\alpha)\Lambda(p)(\bar{V}_\beta+\bar{A}_\beta)\Lambda(p')\right]\delta(p'+q'-p-q)\frac{dq'}{q_0'}\frac{dp'}{p_0'}. \qquad (15.5)$$

Note that in obtaining this expression we used the relationship

$$\tfrac{1}{2}(1+\gamma_5)^2 = (1+\gamma_5).$$

We have

$$\left.\begin{aligned} \overline{\gamma_\beta(1+\gamma_5)} &= \gamma_4(\gamma_\beta(1+\gamma_5))^+\gamma_4 = \gamma_\beta(1+\gamma_5)\,\delta_\beta, \\ \bar{V}_\beta &= \gamma_4 V_\beta^+\gamma_4 = \left(\gamma_\beta G_V + iP_\beta\frac{1}{2M}F_V\right)\delta_\beta, \\ \bar{A}_\beta &= \gamma_4 A_\beta^+\gamma_4 = \left(\gamma_\beta\gamma_5 G_A - i\gamma_5\varkappa_\beta\frac{1}{2M}F_A\right)\delta_\beta \\ (\delta &= -1 \quad \text{for} \quad \beta = 1, 2, 3 \quad \text{and} \quad \delta_4 = 1). \end{aligned}\right\} \qquad (15.6)$$

We now calculate the traces. Using formulae (13.29)–(13.38) we easily find

$$\operatorname{Tr}\left[\gamma_\alpha(1+\gamma_5)\frac{\hat{q}}{2iq_0}\gamma_\beta(1+\gamma_5)\Lambda(q')\right] = -\frac{1}{4q_0 m}2\operatorname{Tr}\left[\gamma_\alpha(1+\gamma_5)\,\hat{q}\gamma_\beta(\hat{q}'+im)\right]$$
$$= -\frac{2}{q_0 m}\left[q_\alpha q_\beta' - \delta_{\alpha\beta}(qq') + q_\alpha' q_\beta + \varepsilon_{\alpha\beta\varrho\sigma}q_\varrho q_\sigma'\right]. \qquad (15.7)$$

The last term in this expression arises from the interference of the vector with the pseudovector. Comparing (15.3) and (14.6) we conclude that $\operatorname{Tr}[V_\alpha\Lambda(p)\,V_\beta\Lambda(p')]$ coincides with $\operatorname{Tr}[R_\alpha\Lambda(p)\,R_\beta\Lambda(p')]$ [expression (14.17)] which was calculated in the preceding section, if in the latter we replace G_M by G_V and F_2 by F_V.

As in the case of electron–nucleon scattering, instead of the form factor F_V we introduce the form factor

$$H_V = G_V - \left(1+\frac{\varkappa^2}{4M^2}\right)F_V. \qquad (15.8)$$

Then

$$\mathrm{Tr}\,[V_\alpha \Lambda(p)\, V_\beta \Lambda(p')] = -\frac{1}{2M^2}\left\{ G_V^2(\delta_{\alpha\beta}\varkappa^2 - \varkappa_\alpha\varkappa_\beta) + \frac{P_\alpha P_\beta}{\varkappa^2+4M^2}(G_V^2\varkappa^2 + 4M^2H_V^2)\right\}. \quad (15.9)$$

Let us examine the trace $\mathrm{Tr}\,[A_\alpha\Lambda(p)\bar{A}_\beta\Lambda(p')]$ (there is no summation over β in this expression).

Using the commutation relations for the γ-matrices we obtain

$$\mathrm{Tr}\,[A_\alpha\Lambda(p)\,\bar{A}_\beta\Lambda(p')\,\delta_\beta] = -\mathrm{Tr}\left[\left(\gamma_\alpha G_A + i\varkappa_\alpha\frac{F_A}{2M}\right)\Lambda(-p)\left(\gamma_\beta G_A + i\varkappa_\beta\frac{F_A}{2M}\right)\Lambda(p')\right], \quad (15.10)$$

where

$$\Lambda(-p) = \frac{-\hat{p}+iM}{2iM}.$$

Comparing this trace with (15.9) we conclude that the trace (15.10) also need not be calculated. This trace can be obtained from (15.9) by the replacement $p \to -p$, $G_V \to G_A$, $F_V \to F_A$ (obviously we do not make the replacement $p \to -p$ in the arguments of the form factors). We obtain

$$\mathrm{Tr}\,[A_\alpha\Lambda(p)\,\bar{A}_\beta\Lambda(p')\,\delta_\beta] = \frac{-1}{2M^2}\left\{G_A^2(-\delta_{\alpha\beta}P^2 + P_\alpha P_\beta) - \frac{\varkappa_\alpha\varkappa_\beta}{P^2+4M^2}(G_A^2P^2 + 4M^2H_A^2)\right\}. \quad (15.11)$$

Here

$$H_A = G_A - \left(1 + \frac{p^2}{4M^2}\right)F_A. \quad (15.12)$$

Thus, if in place of the form factors G_A and F_A we introduce the form factors G_A and H_A, the interference term G_AH_A does not enter the expression for the cross section.

The calculation of the remaining traces does not present any difficulty:

$$\left.\begin{aligned}
\mathrm{Tr}\,[V_\alpha\Lambda(p)\,\bar{A}_\beta\Lambda(p')\,\delta_\beta] &= -\frac{1}{4M^2}\mathrm{Tr}\,[\gamma_\alpha\hat{p}\gamma_\beta\gamma_5\hat{p}'G_VG_A] = -\frac{1}{M^2}\varepsilon_{\alpha\beta\varrho\sigma}p_\varrho p'_\sigma G_VG_A,\\
\mathrm{Tr}\,[A_\alpha\Lambda(p)\,\bar{V}_\beta\Lambda(p')\,\delta_\beta] &= -\frac{1}{M^2}\varepsilon_{\alpha\beta\varrho\sigma}p_\varrho p'_\sigma G_VG_A.
\end{aligned}\right\} \quad (15.13)$$

Now we must multiply (15.7) by the sum of expressions (15.9), (15.11), and (15.13) and sum over α and β. Denote the expression in square brackets on the right-hand side of (15.7) by

$$A_{\alpha\beta} = q_\alpha q'_\beta + q'_\alpha q_\beta - \delta_{\alpha\beta}(qq') + \varepsilon_{\alpha\beta\varrho\sigma}q_\varrho q'_\sigma. \quad (15.14)$$

We first calculate the contraction of $A_{\alpha\beta}$ with the tensors in (15.9), (15.11), and (15.13).

We find

$$\left.\begin{aligned}
A_{\alpha\beta}P_\alpha P_\beta &= 2(Pq)(Pq') - P^2(qq'),\\
A_{\alpha\beta}\delta_{\alpha\beta} &= -2(qq'),\\
A_{\alpha\beta}\varkappa_\alpha\varkappa_\beta &= -m^2(qq'),\\
A_{\alpha\beta}\varepsilon_{\alpha\beta\varrho\sigma}p_\varrho p'_\sigma &= 2[(pq)(p'q') - (pq')(p'q)].
\end{aligned}\right\} \quad (15.15)$$

In obtaining the last contraction we used the relationship

$$\varepsilon_{\alpha\beta\varrho'\sigma'}\varepsilon_{\alpha\beta\varrho\sigma} = 2(\delta_{\varrho'\varrho}\delta_{\sigma'\sigma} - \delta_{\varrho'\sigma}\delta_{\sigma'\varrho}). \tag{15.16}$$

Note also that

$$(Pq') = (P(q+q'-q)) = (Pq)+(P(p-p')) = (Pq). \tag{15.17}$$

From the four-momenta p, q, p', and q' connected by the law of energy-momentum conservation one can construct two independent scalars. We shall choose as independent variables

$$\left.\begin{aligned} s &= -(p+q)^2 = -(p'+q')^2, \\ t &= -(p-p')^2 = -(q'-q)^2 = -\varkappa^2. \end{aligned}\right\} \tag{15.18}$$

It is obvious that in the c.m. system the scalar s is equal to the square of the total energy. We express the scalar products on the right-hand side of (15.15) in terms of s and t. We obtain

$$\left.\begin{aligned} A_{\alpha\beta}P_\alpha P_\beta &= 2[(s-M^2)^2+(t-m^2)(s-\tfrac{1}{4}m^2)], \\ A_{\alpha\beta}\delta_{\alpha\beta} &= -(t-m^2), \\ A_{\alpha\beta}\varkappa_\alpha\varkappa_\beta &= -\tfrac{1}{2}m^2(t-m^2), \\ A_{\alpha\beta}\varepsilon_{\alpha\beta\varrho\sigma}p_\varrho p'_\sigma &= -\tfrac{1}{2}t[2(s-M^2)+(t-m^2)]. \end{aligned}\right\} \tag{15.19}$$

With the help of these relationships it is not difficult to find from (15.9), (15.11), and (15.13) the product of the traces in expression (15.5).

The differential cross section of the process is written in the form

$$d\sigma = B\,\delta(p'+q'-p-q)\,\frac{d\boldsymbol{p}'}{p'_0}\,\frac{d\boldsymbol{q}'}{q'_0}, \tag{15.20}$$

where B is a function of the scalars s and t which can easily be found from (15.5), (15.7), (15.9), (15.11), (15.13), and (15.19).

In order to obtain the final expression for the differential cross section we must integrate (15.20) over four variables (due to the laws of energy-momentum conservation). We perform the integration in the c.m. system. Let us integrate first over $\boldsymbol{p}'$. We obtain

$$d\sigma = B\,\delta(p'_0+q'_0-p_0-q_0)\,\frac{|\boldsymbol{q}'|\,dq'_0\,d\Omega'}{p'_0}. \tag{15.21}$$

The argument of the δ-function is

$$F(q'_0) = q'_0+\sqrt{M^2+q'^2_0-m^2}-p_0-q_0. \tag{15.22}$$

In order to integrate over q' we use formula (14.27). The derivative at the point where $F(q'_0)$ vanishes is

$$\frac{dF(q'_0)}{dq'_0} = \frac{p_0+q_0}{p'_0}. \tag{15.23}$$

As a result we obtain

$$d\sigma = B\,\frac{|\boldsymbol{q}'|\,d\Omega'}{(p_0+q_0)}. \tag{15.24}$$

The scalars s and t are expressed in terms of quantities in the c.m. system in the following manner:

$$s = (p_0+q_0)^2, \quad t = m^2-2q_0q'_0+2q_0|\boldsymbol{q}'|\cos\theta \tag{15.25}$$

(θ is the angle between $\boldsymbol{q}$ and $\boldsymbol{q}'$). The element of the solid angle is

$$d\Omega' = \sin\theta\, d\theta\, d\phi,$$

where ϕ is the azimuth of the vector $\boldsymbol{q}'$. From (15.25) we find that

$$dt = -2q_0|\boldsymbol{q}'|\sin\theta\, d\theta. \tag{15.26}$$

Integrating over ϕ (B is a function of s and t, i.e. B does not depend on ϕ), we obtain the following expression for the cross section:

$$d\sigma = \pi B\frac{dt}{(p_0+q_0)q_0}. \tag{15.27}$$

The cross section $d\sigma$ is a scalar. We must therefore construct a scalar which is equal to $(p_0+q_0)q_0$ in the c.m. system. Obviously

$$(p_0+q_0)\,q_0 = -(p+q)\,q = -(pq).$$

Thus, we find

$$\frac{d\sigma}{dt} = \pi B\frac{1}{-(pq)}. \tag{15.28}$$

Inserting here the expression for B and noting that $-(pq) = \frac{1}{2}(s-M^2)$ we ultimately obtain

$$\frac{d\sigma}{dt} = \frac{G^2}{2\pi}\frac{1}{(s-M^2)^2}\Bigg\{[(s-M^2)^2+(t-m^2)s]\left[\frac{H_V^2-\dfrac{t}{4M^2}G_V^2}{1-\dfrac{t}{4M^2}}+G_A^2\right] + \tfrac{1}{4}m^2(t-m^2)\left[\frac{G_V^2-H_V^2}{1-\dfrac{t}{4M^2}}-\frac{G_A^2-H_A^2}{\dfrac{t}{4M^2}}\right]-t\,[2(s-M^2) + (t-m^2)]\,G_VG_A+\tfrac{1}{2}(t-m^2)\,[G_V^2t+G_A^2(t-4M^2)]\Bigg\}. \tag{15.29}$$

If we make the conserved vector current hypothesis, then

$$G_V = G_M^p-G_M^n, \quad H_V = G_E^p-G_E^n, \tag{15.30}$$

where G_M^p and G_M^n (G_E^p and G_E^n) are the magnetic (charge) form factors of the proton and neutron. These quantities can be found experimentally by scattering electrons on nucleons. The study of the processes (15.1) consequently makes it possible to determine the axial form factors of the nucleon.

We now examine the process

$$\bar{\nu}_\mu + p \rightarrow n + \mu^+ .$$

Denote by q and q' the four-momenta of the antineutrino and μ^+-meson and by p and p' the four-momenta of the proton and neutron. The matrix element of the process is [see (12.126) and (12.127)]

$$(\Phi^+_{q',p'} S\Phi_{q,p}) = i\frac{1}{(2\pi)^6}\left(\frac{M^2 m}{p_0 p'_0 q'_0}\right)^{1/2}\frac{G}{\sqrt{2}}(2\pi)^4$$
$$\times (\bar{u}(-q)\gamma_\alpha(1+\gamma_5)u(-q'))(\bar{u}(p')[V_\alpha(p',p)+A_\alpha(p',p)]u(p))\,\delta(p'+q'-p-q), \quad (15.31)$$

where the matrices $V_\alpha(p', p)$ and $A_\alpha(p', p)$ are given by the expressions (15.3) with the same form factors as for the process (15.1a).

It is obvious that the cross section of the process under consideration can be obtained from expression (15.5) if the lepton trace is replaced by

$$\mathrm{Tr}\left[\gamma_\alpha(1+\gamma_5)\Lambda(-q')\overline{(\gamma_\beta(1+\gamma_5))}\left(\frac{-\hat{q}}{2iq_0}\right)\right]$$
$$= \frac{-2}{mq_0}\,\delta_\beta[q'_\alpha q_\beta - \delta_{\alpha\beta}q'q + q_\alpha q'_\beta - \varepsilon_{\alpha\beta\varrho\sigma}q_\varrho q'_\sigma]. \quad (15.32)$$

Expression (15.32) differs from (15.7) only in the sign of the last term. This term contributes to the cross section only when multiplied by the pseudotensor arising from the interference of the vector and pseudovector [expression (15.13)]. Thus, the cross section of the process

$$\bar{\nu}_\mu + p \rightarrow \mu^+ + n$$

is given by expression (15.29) but with the sign of the term proportional to $G_V G_A$ changed.

16. The Scattering of π-mesons by Nucleons

Here we examine the scattering of π-mesons by nucleons:

$$\pi + p \rightarrow \pi + p. \quad (16.1)$$

This process is determined by strong interactions to which perturbation theory is not applicable.

We first discuss the interaction Hamiltonian of the nucleon and π-meson fields. Denote by $\phi_0(x)$ the field operator of the neutral π-mesons ($\phi_0^+(x) = \phi_0(x)$) and by $\phi(x)$ the field operator of charged π-mesons; $\psi_p(x)$ and $\psi_n(x)$ are the field operators of the proton–antiproton and the neutron–antineutron. The simplest interaction Hamiltonian of meson and nucleon fields constructed by analogy with the electromagnetic interaction Hamiltonian has the form

$$\mathscr{H}_I(x) = ig_p N(\bar{\psi}_p(x)\gamma_5\psi_p(x))\,\phi_0(x) + ig_n N(\bar{\psi}_n(x)\gamma_5\psi_n(x))\,\phi_0(x)$$
$$+ ig_1 N(\bar{\psi}_n(x)\gamma_5\psi_p(x))\phi(x) + ig_1 N(\bar{\psi}_p(x)\gamma_5\psi_n(x))\,\phi^+(x), \quad (16.2)$$

where g_p, g_n, and g_1 are interaction constants. It is not difficult to verify that $\mathscr{H}_I(x)$ is a Hermitian operator.

The Hamiltonian (16.2) is invariant under the gauge transformation

$$\psi_n' = \psi_n, \quad \psi_p' = e^{-ie\chi}\psi_p, \quad \phi' = e^{ie\chi}\phi, \quad \phi_0' = \phi_0. \tag{16.3}$$

This means that the Hamiltonian (16.2) is constructed so that the charge is conserved, with $\phi^{(+)}$ the annihilation operator of a particle with charge e and $\psi_p^{(+)}$ the annihilation operator of a particle with charge $-e$ (e is the charge of the electron).

From experiment it is known that the strong interaction Hamiltonian is invariant under isotopic transformations. We shall ascertain the form which the Hamiltonian (16.2) takes in the case of isotopic invariance. In this case the proton and neutron are two states of the same particle—the nucleon whose isotopic spin is $\frac{1}{2}$. Let ψ_ξ where ξ takes two values (± 1) denote the operator of the nucleon field. We make the condition that $\psi_1 = \psi_p$ and $\psi_{-1} = \psi_n$. Obviously

$$\bar{\psi}_n\gamma_5\psi_p = \bar{\psi}_{\xi'}\gamma_5(\tau_-)_{\xi'\xi}\psi_\xi = \bar{\psi}\gamma_5\tau_-\psi, \tag{16.4}$$

where $(\tau_-)_{-1,1} = 1$ and all remaining elements of the matrix τ_- are zero, i.e.

$$\tau_- = \begin{pmatrix} 0 & 0 \\ 1 & 0 \end{pmatrix} = \tfrac{1}{2}(\tau_1 - i\tau_2). \tag{16.5}$$

Similarly we have

$$\left.\begin{aligned} \bar{\psi}_p\gamma_5\psi_n &= \bar{\psi}\gamma_5\tau_+\psi, \\ \bar{\psi}_p\gamma_5\psi_p &= \bar{\psi}\gamma_5\,\tfrac{1}{2}(1+\tau_3)\,\psi, \\ \bar{\psi}_n\gamma_5\psi_n &= \bar{\psi}\gamma_5\,\tfrac{1}{2}(1-\tau_3)\,\psi, \end{aligned}\right\} \tag{16.6}$$

where

$$\tau_+ = \tfrac{1}{2}(\tau_1 + i\tau_2). \tag{16.7}$$

In relationships (16.5)–(16.7) τ_1, τ_2, and τ_3 are Pauli matrices.

With the help of (16.4) and (16.6) the interaction Hamiltonian (16.2) can be written in the form

$$\begin{aligned} \mathscr{H}_I = {} & \frac{g_p+g_n}{2} N(\bar{\psi}\gamma_5\psi)\,\phi_3 + \frac{g_p-g_n}{2} N(\bar{\psi}\gamma_5\tau_3\psi)\,\phi_3 \\ & + \frac{g_1}{\sqrt{2}} N(\bar{\psi}\gamma_5\tau_1\psi)\,\phi_1 + \frac{g_1}{\sqrt{2}} N(\bar{\psi}\gamma_5\tau_2\psi)\,\phi_2. \end{aligned} \tag{16.8}$$

Here

$$\phi_1 = \frac{\phi^+ + \phi}{\sqrt{2}}, \quad \phi_2 = \frac{i(\phi^+ - \phi)}{\sqrt{2}}, \quad \phi_3 = \phi_0. \tag{16.9}$$

Note that the operators ϕ_1, ϕ_2, and ϕ_3 are Hermitian. From (16.9) we obtain

$$\phi = \frac{\phi_1 + i\phi_2}{\sqrt{2}}; \quad \phi^+ = \frac{\phi_1 - i\phi_2}{\sqrt{2}}. \tag{16.10}$$

In the framework of the isotopic invariance hypothesis π-, π^--, and π^0-mesons are three states of one particle whose isotopic spin is equal to one. Under isotopic transformations $\bar{\psi}\gamma_5\tau_i\psi$ and ϕ_i are vectors and $\bar{\psi}\gamma_5\psi$ is a scalar. So that the interaction Hamiltonian (16.8) is invariant under isotopic transformations, the coupling constants must satisfy the relationship

$$g_p = -g_n = \frac{g_1}{\sqrt{2}} = g. \tag{16.11}$$

Actually, in this case the first term in (16.8) (the product of a scalar by the third component of a vector) disappears and the three following terms are equal to

$$\mathcal{H}_I = igN(\bar{\psi}\gamma_5\tau_i\psi)\,\phi_i. \tag{16.12}$$

Although perturbation theory is not applicable in this case, nevertheless it is useful to examine the diagrams of the lowest order of perturbation theory in g. First consider the process

$$\pi^+ + p \rightarrow \pi^+ + p. \tag{16.13}$$

We denote by q and q' the four-momenta of the initial and final π^+-mesons and by p and p' the four-momenta of the initial and final protons. Obviously a contribution to the matrix element of the process is made by the following operator:

$$\bar{\psi}_p^{(-)}(x_1)\,\gamma_5\,\overbrace{\psi_n(x_1)\,\bar{\psi}_n(x_2)}\,\gamma_5\psi_p^{(+)}(x_2)\,N\big(\phi^{+(+)}(x_1)\,\phi^{(-)}(x_2)\big). \tag{16.14}$$

Calculating the matrix element we obtain

$$(\Phi^+_{q',p'}S^{(2)}\Phi_{q,p}) = (-i)^2\frac{1}{(2\pi)^6}\left(\frac{M^2}{4q_0q_0'p_0p_0'}\right)^{1/2}\frac{(-1)}{(2\pi)^4}$$
$$\times(2\pi)^8(ig_1)^2\left(\bar{u}(p')\,\gamma_5\frac{1}{\hat{p}-\hat{q}'-iM}\gamma_5 u(p)\right)\delta(p'+q'-p-q). \tag{16.15}$$

This matrix element is described by the diagram in Fig. 22. The external meson line with momentum q is put into correspondence with $\left(1/(2\pi)^{3/2}\right)\left(1/\sqrt{2q_0}\right)$ and the vertex is put into correspondence with $(ig_1)\gamma_5(2\pi)^4\,\delta(P'-P)$ where P' and P are the total four-momenta of the emerging and entering particles. For the external and internal solid lines we assume the same rules of correspondence as for the electron. The diagram in Fig. 22 describes the

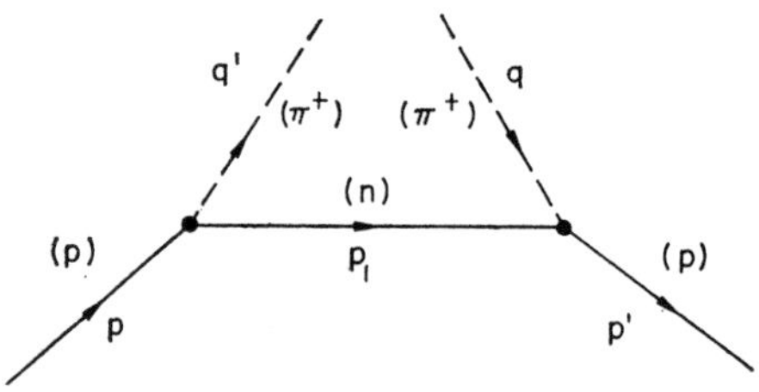

FIG. 22

following sequence of absorption and emission events: the initial proton emits the final π^+-meson and turns into a virtual neutron; then the virtual neutron absorbs the initial π^+-meson and changes into the proton in the final state. The diagram in Fig. 22 is equivalent to one of the diagrams for the Compton effect (see Fig. 7). It is obvious that the second diagram (absorption of the initial π^+-meson and so forth) is forbidden by the law of charge conservation (in the system under discussion there are no particles with charge $2e$). The scattering of π^--mesons by protons in the second order in g is described by the diagram in Fig. 23.

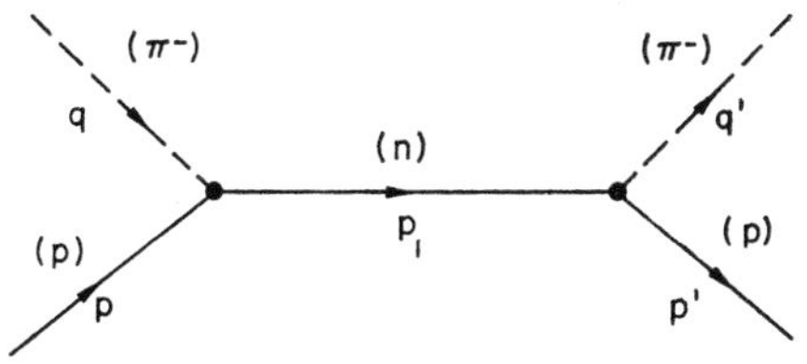

FIG. 23

Thus, the rules for constructing diagrams which describe the processes of meson–nucleon interactions are the same as in quantum electrodynamics. The law of charge conservation in the case of charged particles, however, leads to a reduction in the overall number of diagrams in a given order of perturbation theory.

Now we shall construct a general expression for the matrix element of the process of π-meson–nucleon scattering on the basis of invariance principles. Then we shall calculate the cross section for the scattering of π-mesons by polarized nucleons.

The matrix element of the process (16.1) is written in the form

$$(\Phi^+_{q',p'}(S-1)\Phi_{q,p}) = \frac{1}{(2\pi)^2}\frac{1}{\sqrt{p_0p_0'q_0q_0'}}M_{q',p';\,q,p}\,\delta(p'+q'-p-q). \tag{16.16}$$

Here p and q (p' and q') are the four-momenta of the initial (final) nucleon and meson. The matrix element $M_{q',p';\,q,p}$ has the form

$$M_{q',p';\,q,p} = \bar{u}(p')\,\mathfrak{M}(q',p';q,p)\,u(p) \tag{16.17}$$

and is a scalar (the interaction conserves parity). The cross section for the process is given by expression (13.26).

Let us construct the general expression for the matrix $\mathfrak{M}(q',p';q,p)$. The matrix $\mathfrak{M}$ is a 4×4 matrix and can always be expanded into sixteen Dirac matrices:

$$\mathfrak{M}(q',p';q,p) = A+B_\mu\gamma_\mu+C_{\mu\nu}\sigma_{\mu\nu}+D_\mu\gamma_\mu\gamma_5+E\gamma_5, \tag{16.18}$$

where the coefficients of the expansion depend on the four-vectors $p, p', \ldots$. Since the vectors p, q, p', and q' are related by the law of energy-momentum conservation

$$p+q = p'+q', \tag{16.19}$$

only three of them are independent. We choose as independent vectors

$$\left.\begin{aligned} P &= p+p', \\ \varkappa &= p'-p = q-q', \\ Q &= q+q'. \end{aligned}\right\} \tag{16.20}$$

The first term in the expansion (16.18) is scalar. From the four vectors p, p', q, and q' obeying the conservation law (16.19) and the relations $p^2 = p'^2 = -M^2$, $q^2 = q'^2 = -m_\pi^2$ (M is the mass of the nucleon, m_π is the mass of the π-meson) it is possible to produce two independent scalars. We choose as independent variables the scalars

$$\left.\begin{aligned} s &= -(p+q)^2, \\ t &= -(p-p')^2 = -\varkappa^2. \end{aligned}\right\} \tag{16.21}$$

Thus, A is a function of s and t. The quantity $\bar{u}(p')\gamma_\mu u(p)$ transforms like a vector. Since $\bar{u}(p')\gamma_\mu u(p)B_\mu$ is a scalar, the coefficient B_μ must be a vector. We have the following general expression for B_μ:

$$B_\mu = B_1 P_\mu + B_2 \varkappa_\mu + BQ_\mu, \tag{16.22}$$

where B_1, B_2, and B are functions of the scalar s and t. Considering the Dirac equation

$$\left.\begin{aligned} (\hat{p}-iM)\,u(p) &= 0, \\ \bar{u}(p')(\hat{p}'-iM) &= 0, \end{aligned}\right\} \tag{16.23}$$

it is not difficult to show that the second term in (16.22) does not contribute to the matrix element (16.17) and the first term in (16.22) reduces to term A. Actually,

$$\left.\begin{aligned} \bar{u}(p')(\hat{p}'+\hat{p})\,u(p) &= 2iM\bar{u}(p')\,u(p), \\ \bar{u}(p')(\hat{p}'-\hat{p})\,u(p) &= 0. \end{aligned}\right\} \tag{16.24}$$

Thus, in (16.22) we need take into account only the last term BQ_μ.

Obviously $C_{\mu\nu}$ is an antisymmetric second-rank tensor having the following general form:

$$C_{\mu\nu} = C_1(P_\mu Q_\nu - P_\nu Q_\mu) + C_2(P_\mu \varkappa_\nu - P_\nu \varkappa_\mu) + C_3(Q_\mu \varkappa_\nu - Q_\nu \varkappa_\mu), \tag{16.25}$$

where C_1, C_2, and C_3 are functions of the scalars s and t. Using equations (16.23) it is easy to show that in the matrix element all terms of (16.25) reduce to A and BQ_μ. Actually,

$$\bar{u}(p')\,\sigma_{\mu\nu}(Q_\mu\varkappa_\nu - Q_\nu\varkappa_\mu)\,u(p) = \frac{1}{i}\,2\bar{u}(p')\,\hat{Q}\hat{\varkappa}u(p) = 4\frac{1}{i}\,\bar{u}(p')[(p'Q)-iM\hat{Q}]\,u(p)$$

and so forth.

It is clear that the coefficient D_μ must transform like a pseudovector. We obtain the following general expression for D_μ:

$$D_\mu = D\varepsilon_{\mu\nu\varrho\sigma}p_\nu p'_\varrho Q_\sigma. \tag{16.26}$$

Using relationship (12.42) it is not difficult to show that this term in the matrix element also reduces to A and BQ_μ. Finally, the last coefficient in the expansion (16.18) must be a pseudoscalar. A pseudoscalar cannot be constructed from three four-vectors. Consequently, $E = 0$.

We finally come to the conclusion that the matrix $\mathfrak{M}(q', p'; q, p)$ has the following general form:

$$\mathfrak{M}(q', p'; q, p) = a + b\hat{Q}, \tag{16.27}$$

where a and b are functions of the scalars s and t. These functions are defined by the dynamics of the process.

With the help of (13.26) we obtain for the cross section of π-meson–nucleon scattering

$$d\sigma = \frac{1}{(2\pi)^2} \frac{1}{\sqrt{(pq)^2 - M^2 m_\pi^2}} \operatorname{Tr}\,[\mathfrak{M}\varrho(p)\,\overline{\mathfrak{M}}\Lambda(p')]\;\delta(p'+q'-p-q)\frac{dp'}{p_0'}\frac{dq'}{q_0'}. \tag{16.28}$$

We shall suppose that the initial nucleons are polarized. The spin density matrix of the initial state is equal to (see Appendix)

$$\varrho(p) = \Lambda(p)\tfrac{1}{2}(1 + i\gamma_5\hat{\xi}), \tag{16.29}$$

where ξ_μ is the polarization four-vector and

$$\Lambda(p) = \frac{(\hat{p} + iM)}{2iM}$$

is a projection operator. The vector ξ_μ satisfies the condition

$$\xi p = 0. \tag{16.30}$$

We shall calculate the trace in (16.28). Considering that

$$\bar{\hat{Q}} = \gamma_4 \hat{Q}^+ \gamma_4 = -\hat{Q},$$

we obtain

$$\operatorname{Tr}\,[\mathfrak{M}\varrho(p)\,\overline{\mathfrak{M}}\Lambda(p')] = \frac{-1}{8M^2}\operatorname{Tr}\,[(a+b\hat{Q})(\hat{p}+iM)(1+i\gamma_5\hat{\xi})(a^* - b^*\hat{Q})(\hat{p}' - iM)]$$

$$= -\frac{1}{2M^2}[|a|^2(pp' - M^2) + 2\operatorname{Im} ab^* M(PQ) - |b|^2(2(Qp)(Qp') - Q^2(pp' + M^2))$$

$$+ 2i \operatorname{Re} ab^* \varepsilon_{\mu\nu\varrho\sigma}\xi_\mu p_\nu p_\varrho' Q_\sigma]. \tag{16.31}$$

We express the scalar products in this expression in terms of s, t, and

$$u = -(p - q')^2. \tag{16.32}$$

The variables s, t, and u are related by

$$s + t + u = 2M^2 + 2m_\pi^2, \tag{16.33}$$

which can be obtained easily from the law of energy-momentum conservation.

We have

$$\operatorname{Tr}\,[\mathfrak{M}\varrho\overline{\mathfrak{M}}\Lambda(p')] = \frac{1}{4M^2}[|a|^2(4M^2 - t) + 4M \operatorname{Im} ab^*(s-u)$$

$$+ |b|^2((s-u)^2 - t(t - 4m_\pi^2)) + 4 \operatorname{Re} ab^* N_\mu \xi_\mu]. \tag{16.34}$$

Here

$$N_\mu = -i\varepsilon_{\mu\nu\varrho\sigma} p_\nu p'_\varrho Q_\sigma. \tag{16.35a}$$

Obviously the pseudovector N_μ can also be written in the following manner:

$$N_\mu = -2i\varepsilon_{\mu\nu\varrho\sigma} p_\nu p'_\varrho (p+q)_\sigma. \tag{16.35b}$$

Let us calculate the cross section for the process of π-meson–nucleon scattering in the c.m. system. We integrate (16.28) first over $\boldsymbol{p}'$ and then over q'_0. From (14.27) we find

$$\delta(p'_0+q'_0-p_0-q_0) = \frac{\delta(q'_0-q_0)}{\dfrac{p_0+q_0}{p_0}}. \tag{16.36}$$

Furthermore, with the help of (13.18) we find that in the c.m. system

$$\sqrt{(pq)^2-M^2m_\pi^2} = |\boldsymbol{q}|(p_0+q_0).$$

For the differential cross section of the process we have the expression

$$\frac{d\sigma}{d\Omega} = \frac{1}{(2\pi)^2}\frac{1}{(p_0+q_0)^2}\,\mathrm{Tr}\,[\mathfrak{M}\varrho(p)\,\overline{\mathfrak{M}}\Lambda(p')], \tag{16.37}$$

where the trace is given by expression (16.34). The variables s and t are expressed in the following manner by quantities in the c.m. system:

$$\left.\begin{aligned} s &= (p_0+q_0)^2, \\ t &= -(\boldsymbol{q}-\boldsymbol{q}')^2 = -2|\boldsymbol{q}|^2(1-\cos\theta). \end{aligned}\right\} \tag{16.38}$$

From (16.35b) it is obvious that in the c.m. system

$$N_4 = 0, \quad N_k = 2(p_0+q_0)(\boldsymbol{q}\times\boldsymbol{q}')_k. \tag{16.39}$$

It is not difficult to see that the cross section for π-meson–nucleon scattering in the c.m. system can be written

$$\frac{d\sigma}{d\Omega} = \left(\frac{d\sigma}{d\Omega}\right)_0 (1+\boldsymbol{\xi}\boldsymbol{A}^0). \tag{16.40}$$

Here

$$\left(\frac{d\sigma}{d\Omega}\right)_0 = \frac{1}{4(2\pi)^2}\frac{1}{sM^2}\,[|a|^2(4M^2-t)+4M\,\mathrm{Im}\,ab^*(s-u)+|b|^2((s-u)^2-t(t-4m_\pi^2))] \tag{16.41}$$

is the cross section for the scattering of mesons by unpolarized nucleons and

$$\boldsymbol{A}^0 = \frac{2(\boldsymbol{q}\times\boldsymbol{q}')\,\mathrm{Re}\,ab^*}{(2\pi)^2(p_0+q_0)M^2\left(\dfrac{d\sigma}{d\Omega}\right)_0}. \tag{16.42}$$

The pseudovector $\boldsymbol{A}^0$ coincides with the polarization of the final nucleon which arises from the scattering of π-mesons on unpolarized nucleons. Actually, the polarization four-vector of the final nucleons when the initial nucleons are unpolarized is equal to

$$\xi_\mu^0 = \frac{\frac{1}{2}\,\mathrm{Tr}\,[i\gamma_5\gamma_\mu\Lambda(p')\,\mathfrak{M}\Lambda(p)\,\overline{\mathfrak{M}}\Lambda(p')]}{\frac{1}{2}\,\mathrm{Tr}\,[\mathfrak{M}\Lambda(p)\,\overline{\mathfrak{M}}\Lambda(p')]}. \tag{16.43}$$

Using the transformation relations for the matrices γ_μ we obtain

$$i\gamma_5\gamma_\mu\Lambda(p') = \frac{p'_\mu}{M}\gamma_5 + \Lambda(p')\,i\gamma_5\gamma_\mu. \tag{16.44}$$

It is obvious that the term $(P'_\mu/M)\gamma_5$ does not contribute to (16.43) ($\mathrm{Tr}\,[\gamma_5\mathfrak{M}\Lambda(p)\,\overline{\mathfrak{M}}\Lambda(p)]$ is a pseudoscalar; since it is impossible to construct a pseudoscalar from the three independent vectors, this trace is zero). Considering that $\Lambda(p')\,\Lambda(p') = \Lambda(p')$ we obtain

$$\xi_\mu^0 = \frac{\frac{1}{2}\,\mathrm{Tr}\,[i\gamma_5\gamma_\mu\,\mathfrak{M}\Lambda(p)\,\overline{\mathfrak{M}}\Lambda(p')]}{\frac{1}{2}\,\mathrm{Tr}\,[\mathfrak{M}\Lambda(p)\,\overline{\mathfrak{M}}\Lambda(p')]}. \tag{16.45}$$

Calculating the trace in the numerator of this expression we find

$$\xi_\mu^0 = \frac{\dfrac{1}{M^2}\,\mathrm{Re}\,ab^*N_\mu}{\frac{1}{2}\,\mathrm{Tr}\,[\mathfrak{M}\Lambda(p)\,\overline{\mathfrak{M}}\Lambda(p')]}. \tag{16.46}$$

The trace in the denominator is given by expression (16.34) with $\xi_\mu = 0$. In the c.m. system $\xi_4^0 = 0$ and

$$\boldsymbol{\xi}^0 = \boldsymbol{A}^0. \tag{16.47}$$

The vector $\boldsymbol{A}^0$ is called the asymmetry vector. We have verified that the vectors of polarization and asymmetry are equal for the elastic scattering of particles with spin zero on particles with spin $\frac{1}{2}$. Note that this equality is a result of the general principles of invariance and is valid for elastic scattering of particles with arbitrary spins.

Finally, we shall make the connection with expression (16.40). The polarization vector of a particle in the system where its momentum is $\boldsymbol{p}$ is related to the polarization vector $\boldsymbol{\xi}_r$ in the rest system by the relationship (see Appendix)

$$\boldsymbol{\xi} = \boldsymbol{\xi}_r + \frac{(\boldsymbol{\xi}_r\boldsymbol{p})\boldsymbol{p}}{(p_0+M)M}. \tag{16.48}$$

Since the vector $\boldsymbol{\xi}^0$ is directed along the normal to the scattering plane, then from (16.48) it is obvious that

$$\boldsymbol{\xi}\boldsymbol{\xi}^0 = \boldsymbol{\xi}_r\boldsymbol{\xi}^0. \tag{16.49}$$

From (16.48) it also follows that

$$\boldsymbol{\xi}^0 = \boldsymbol{\xi}_r^0.$$

Thus,

$$\boldsymbol{\xi}\boldsymbol{\xi}^0 = \boldsymbol{\xi}_r\boldsymbol{\xi}_r^0. \tag{16.50}$$

17. The Decay $\Lambda \to p+l^- + \bar{\nu}$

In this section we examine the decay

$$\Lambda \to p+l^- + \bar{\nu}, \tag{17.1}$$

where l is the charged lepton (electron or μ-meson).

The Hamiltonian has the form (12.79). In the first order of the weak interaction constant the decay matrix element is

$$(\Phi^+_{q,q',p'}S\Phi_p) = -i\frac{G}{\sqrt{2}}\frac{1}{(2\pi)^3}\left(\frac{m_l}{q_0}\right)^{1/2}(\bar{u}(q)\,\gamma_\alpha(1+\gamma_5)\,u(-q'))$$
$$\times(\Phi^+_{p'}J_\alpha(0)\,\Phi_p)\,(2\pi)^4\,\delta(p'+q'+q-p). \tag{17.2}$$

Here q and q' are the four-momenta of the lepton and the antineutrino; p and p' are the four-momenta of the initial and final fermions; m_l is the mass of the lepton and $J_\alpha = V_\alpha + A_\alpha$ is the sum of the vector and axial currents in the Heisenberg representation (with allowance for strong interactions).

The matrix element $(\Phi^+_{p'}J_\alpha(0)\,\Phi_p)$ can be written in the form

$$(\Phi^+_{p'}J_\alpha(0)\,\Phi_p) = \frac{1}{(2\pi)^3}\left(\frac{M_\Lambda M_p}{p_0 p_0'}\right)^{1/2}\bar{u}(p')\,[V_\alpha(p',p)+A_\alpha(p',p)]\,u(p), \tag{17.3}$$

where M_Λ and M_p are respectively the masses of the Λ-hyperon and proton.

Lorentz invariance requires that the matrices $V_\alpha(p',p)$ and $A_\alpha(p',p)$ be characterized by three form factors and have the form (12.90). Using the relationship

$$\begin{gathered} i\bar{u}(p')\,u(p)\,P_\alpha = -\bar{u}(p')\,\sigma_{\alpha\beta}\varkappa_\beta\,u(p) - (M_\Lambda + M_p)\,\bar{u}(p')\,\gamma_\alpha u(p) \\ (P = p+p', \quad \varkappa = p'-p), \end{gathered} \tag{17.4}$$

we write the matrix element $\bar{u}(p')\,V_\alpha(p',p)\,u(p)$ in the following manner:

$$\bar{u}(p')\,V_\alpha(p',p)\,u(p) = \bar{u}(p')\left[\gamma_\alpha\left(G_V - \frac{M_\Lambda + M_p}{2M_\Lambda}F_V\right) - \frac{F_V}{2M_\Lambda}\sigma_{\alpha\beta}\varkappa_\beta + i\varkappa_\alpha\frac{H_V}{2M_\Lambda}\right]u(p). \tag{17.5}$$

The second and third terms of this expression are proportional to the four-momentum transfer. If all form factors are of the same order, the contribution of these terms is less than or of the order of $(M_\Lambda - M_P)/M_\Lambda$ times the contribution of the first term. Thus, it is natural to expect that the basic contribution to the decay matrix element is made by the first term on the right-hand side of (17.5). We can also conclude from the relationship

$$i\bar{u}(p')\,\gamma_5 u(p) = \frac{1}{(M_\Lambda + M_p)}\bar{u}(p')\,\hat{\varkappa}\gamma_5 u(p) \tag{17.6}$$

that the contributions of the second and third terms of the expression for $A_\alpha(p',p)$ [see (12.90)] are small in comparison to the contribution of the first term.

We shall consider only those terms which make a substantial contribution to the matrix element of the process. Disregarding in this approximation the dependence of the form factors on $\varkappa^2$, we obtain the following expression for the matrix element of the decay:

$$(\Phi^{+}_{q,\,q',\,p'}S\Phi) = \frac{1}{(2\pi)^2}\frac{1}{\sqrt{p_0 p_0' q_0 q_0'}} M_{q,\,q',\,p';\,p}\,\delta(p'+q'+q-p), \tag{17.7}$$

where

$$M_{q,\,q',\,p';\,p} = -i\frac{G}{\sqrt{2}}\sqrt{m_l M_\Lambda M_p q_0'}(\bar{u}(q)\,\gamma_\alpha(1+\gamma_5)\,u(-q'))$$
$$\times(\bar{u}(p')\,\gamma_\alpha(g_V+g_A\gamma_5)\,u(p)) \tag{17.8}$$

(g_V and g_A are constants).

The decay probability per unit time is

$$d\Gamma = \frac{dw}{\varrho} = \frac{1}{(2\pi)^5}\frac{1}{p_0}\sum|M_{q,\,q',\,p';\,p}|^2\,\delta(p'+q'+q-p)\frac{dp'}{p_0'}\frac{dq}{q_0}\frac{dq'}{q_0'}. \tag{17.9}$$

Here dw is the decay probability per unit volume and unit time; $\varrho = 1/(2\pi)^3$ is the initial density and $\sum$ denotes the summation over the spin states of the final particles and the averaging over the spin states of the initial particle. From (17.8) and (17.9) we obtain for the decay probability

$$d\Gamma = \frac{1}{(2\pi)^5}\frac{1}{p_0}\frac{G^2}{2^5}\,\mathrm{Tr}\,[\gamma_\alpha(1+\gamma_5)\,\hat{q}'\gamma_\beta(1+\gamma_5)\,(\hat{q}+im_l)]$$
$$\times\mathrm{Tr}\left[\gamma_\alpha(g_V+g_A\gamma_5)\,(\hat{p}+iM_\Lambda)\frac{(1+i\gamma_5\hat{\xi})}{2}\gamma_\beta(g_V^*+g_A^*\gamma_5)\right.$$
$$\left.\times(\hat{p}'+iM_p)\right]\delta(p'+q'+q-p)\frac{dp'}{p_0'}\frac{dq'}{q_0'}\frac{dq}{q_0}. \tag{17.10}$$

Here ξ_μ is the polarization four-vector of the initial particles.

Let us examine the decay of unpolarized Λ-hyperons. The calculation of the traces does not present any difficulties. We obtain

$$\mathrm{Tr}\,[\gamma_\alpha(g_V+g_A\gamma_5)\,(\hat{p}+iM_\Lambda)\,\gamma_\beta(g_V^*+g_A^*\gamma_5)\,(\hat{p}'+iM_p)]$$
$$= 4[(|g_V|^2+|g_A|^2)\,(p_\alpha p_\beta'-\delta_{\alpha\beta}pp'+p_\alpha' p_\beta)$$
$$-(|g_V|^2-|g_A|^2)\,\delta_{\alpha\beta}M_\Lambda M_p+2\mathrm{Re}\,g_V g_A^*\varepsilon_{\alpha\beta\varrho\sigma}p_\varrho p_\sigma']. \tag{17.11}$$

The second trace can be obtained from (17.11) if we set $g_V = g_A = 1$ and make the replacement $M_\Lambda \to 0$, $M_p \to m_1$, $p \to q'$, $p' \to q$. We find

$$\mathrm{Tr}\,[\gamma_\alpha(1+\gamma_5)\,\hat{q}'\gamma_\beta(1+\gamma_5)(\hat{q}+im_l)] = 8[q_\alpha' q_\beta-\delta_{\alpha\beta}q'q+q_\alpha q_\beta'+\varepsilon_{\alpha\beta\varrho\sigma}q_\varrho' q_\sigma]. \tag{17.12}$$

Multiplying these expressions and summing over α and β we obtain

$$d\Gamma = \frac{1}{(2\pi)^5}\frac{1}{p_0}G^2[(|g_V|^2+|g_A|^2)((pq')(p'q)+(pq)(p'q'))+(|g_V|^2-|g_A|^2)\,M_\Lambda M_p(qq')$$
$$+2\mathrm{Re}\,g_V g_A^*((pq')(p'q)-(pq)(p'q'))]\,\delta(p'+q'+q-p)\frac{dp'}{p_0'}\frac{dq'}{q_0'}\frac{dq}{q_0}. \tag{17.13}$$

Let us calculate the spectrum of the decay nucleons. For this we integrate (17.13) over $\boldsymbol{q}'$ and $\boldsymbol{q}$. We introduce the four-vector

$$k = p-p' = q+q' \tag{17.14}$$

and we replace $q \to k-q'$ in the square brackets of expression (17.13). We obtain

$$\begin{aligned} d\Gamma = \frac{1}{(2\pi)^5}\frac{1}{p_0}G^2[(|g_V|^2+|g_A|^2)((pq')(p'k)+(pk)(p'q')-2(pq')(p'q')) \\ +(|g_V|^2-|g_A|^2)M_AM_p(kq')+2\operatorname{Re} g_Vg_A^*((pq')(p'k)-(pk)(p'q'))] \\ \times\delta(q+q'-k)\frac{d\boldsymbol{p}'}{p_0'}\frac{d\boldsymbol{q}'}{q_0'}\frac{d\boldsymbol{q}}{q_0}. \end{aligned} \tag{17.15}$$

From (17.14) it follows that

$$q^2 = (k-q')^2 = -m_l^2.$$

From this we obtain

$$kq' = \tfrac{1}{2}(k^2+m_l^2). \tag{17.16}$$

When this relationship for the decay probability is integrated over the momenta of the lepton and neutrino, we find

$$\begin{aligned} d\Gamma(p') = \frac{1}{(2\pi)^5}\frac{1}{p_0}G^2[(|g_V|^2+|g_A|^2)((p_\alpha I_\alpha)(p'k)+(pk)(p_\alpha' I_\alpha)-2p_\alpha p_\beta' I_{\alpha\beta}) \\ +(|g_V|^2-|g_A|^2)M_AM_p\tfrac{1}{2}(k^2+m_l^2)I+2\operatorname{Re} g_Vg_A^*((p_\alpha I_\alpha)(p'k)-(pk)(p_\alpha' I_\alpha))]\frac{d\boldsymbol{p}'}{p_0'}. \end{aligned} \tag{17.17}$$

Here

$$I = \int \frac{d\boldsymbol{q}}{q_0}\frac{d\boldsymbol{q}'}{q_0'}\delta(q+q'-k), \tag{17.18}$$

$$I_\alpha = \int q_\alpha' \frac{d\boldsymbol{q}}{q_0}\frac{d\boldsymbol{q}'}{q_0'}\delta(q+q'-k), \tag{17.19}$$

$$I_{\alpha\beta} = \int q_\alpha' q_\beta' \frac{d\boldsymbol{q}}{q_0}\frac{d\boldsymbol{q}'}{q_0'}\delta(q+q'-k). \tag{17.20}$$

We shall first examine the integral (17.18). It is obviously a scalar. Let us calculate this integral in the system where $\boldsymbol{k} = 0$ (the c.m. system of the lepton and neutrino). Integrating over the momentum of the lepton we obtain

$$I = \int \frac{1}{q_0} q_0' \, dq_0' \, d\Omega' \, \delta(q_0+q_0'-k_0), \tag{17.21}$$

where

$$q_0 = \sqrt{m_l^2+q_0'^2}. \tag{17.22}$$

Furthermore, using (14.27) and considering that

$$\frac{d(q_0+q_0')}{dq_0'} = \frac{k_0}{q_0}, \tag{17.23}$$

we find

$$I = 4\pi \frac{q_0'}{k_0}. \tag{17.24}$$

We multiply the numerator and denominator of this expression by k_0. It is obvious that $k_0 q_0'$ and k_0^2 are the invariant quantities $-kq'$ and $-k^2$ in our reference system. We ultimately obtain using (17.16)

$$I = 2\pi \frac{k^2+m_l^2}{k^2}. \tag{17.25}$$

Let us now examine the integral I_α. From considerations of invariance it is clear that

$$I_\alpha = Ak_\alpha, \tag{17.26}$$

where A is a scalar. Multiplying this relationship by k_α and using (17.16) we obtain

$$Ak^2 = \int (q'k) \frac{dq}{q_0} \frac{dq'}{q_0'} \delta(q+q'-k) = \frac{k^2+m_l^2}{2} I. \tag{17.27}$$

Thus,

$$I_\alpha = \frac{k^2+m_l^2}{2k^2} Ik_\alpha. \tag{17.28}$$

Finally, we examine the integral $I_{\alpha\beta}$ defined by expression (17.20). Obviously $I_{\alpha\beta}$ is a second-rank tensor and has the following general form:

$$I_{\alpha\beta} = B\delta_{\alpha\beta} + Ck_\alpha k_\beta, \tag{17.29}$$

where B and C are scalars. Multiplying (17.29) by $\delta_{\alpha\beta}$, summing over α and β, and remembering that $q'^2 = 0$, we obtain

$$0 = 4B + Ck^2. \tag{17.30}$$

We now multiply (17.29) by $k_\alpha k_\beta$. With the help of (17.16) we find

$$\tfrac{1}{4}(k^2+m_l^2)^2 I = Bk^2 + C(k^2)^2. \tag{17.31}$$

From (17.30) and (17.31) we obtain

$$B = -\tfrac{1}{12} \frac{(k^2+m_l^2)^2}{k^2} I; \quad C = \tfrac{1}{3} \frac{(k^2+m_l^2)^2}{(k^2)^2} I. \tag{17.32}$$

Inserting these expressions into (17.29) we find

$$I_{\alpha\beta} = \left(-\tfrac{1}{4}\delta_{\alpha\beta} + \frac{k_\alpha k_\beta}{k^2}\right) \tfrac{1}{3} \frac{(k^2+m_l^2)^2}{k^2} I. \tag{17.33}$$

From (17.26) it is clear that

$$(p_\alpha I_\alpha)(p'k) - (pk)(p_\alpha' I_\alpha) = 0. \tag{17.34}$$

With the help of (17.25), (17.28), and (17.33) we obtain the following expression for the decay probability integrated over the momenta of the lepton and the neutrino:

$$d\Gamma(p') = \frac{1}{(2\pi)^4} G^2 \frac{1}{p_0} \left(\frac{k^2+m_l^2}{k^2}\right)^2 \left\{\tfrac{1}{3}(|g_V|^2+|g_A|^2)\left[\tfrac{1}{2}(pp')(k^2+m_l^2)+(pk)(p'k)\frac{k^2-2m_l^2}{k^2}\right] + \tfrac{1}{2}(|g_V|^2-|g_A|^2)M_\Lambda M_p k^2\right\}\frac{dp'}{p_0'}. \tag{17.35}$$

Let us find the range of variation of k^2. Obviously

$$k^2 = -(\sqrt{m_l^2+q_0'^2}+q_0')^2, \tag{17.36}$$

where q_0' is the energy of the neutrino in the c.m. system of the lepton and neutrino.

Since $q_0' \geqslant 0$, from (17.36) we find

$$k^2 \leqslant -m_l^2. \tag{17.37}$$

In order to obtain the lower limit on the variable k^2, we write k^2 in the rest system of the initial particle:

$$k^2 = -M_\Lambda^2 - M_p^2 - 2pp' = -M_\Lambda^2 - M_p^2 + 2M_\Lambda E, \tag{17.38}$$

where E is the total energy of the proton in the Λ-hyperon rest system. From this we obtain ($E \geqslant M_P$)

$$k^2 \geqslant -(M_\Lambda - M_p)^2. \tag{17.39}$$

Thus,

$$-(M_\Lambda - M_p)^2 \leqslant k^2 \leqslant -m_l^2. \tag{17.40}$$

Let us find the proton recoil spectrum in the Λ-hyperon rest system. From (17.37) and (17.38) we find that the maximum energy of the proton in this system is

$$E_0 = \frac{M_\Lambda^2 + M_p^2 - m_l^2}{2M_\Lambda}. \tag{17.41}$$

From (17.38) and (17.41) we find

$$k^2 + m_l^2 = -2M_\Lambda(E_0 - E). \tag{17.42}$$

We shall give the expression for that part of the proton spectrum where $-k^2 \gg m_l^2$. From (17.35), omitting terms of order m_l^2/k^2 and using (17.42) we obtain

$$\frac{d\Gamma}{dE} = \frac{2}{(2\pi)^2} G^2 M_\Lambda (E^2 - M_p^2)^{1/2} \{(|g_V|^2+|g_A|^2) \times [E(E_0-E)+\tfrac{1}{3}(E^2-M_p^2)] - (|g_V|^2-|g_A|^2) M_p(E_0-E)\}. \tag{17.43}$$

18. The Decay $K^- \to \pi^0 + l^- + \bar{\nu}$

The next example also relates to weak interaction physics. We shall discuss the decay of a charged K^--meson into a π^0-meson, a charged lepton (electron, or μ^--meson), and an antineutrino:

$$K^- \to \pi^0 + l^- + \bar{\nu}. \tag{18.1}$$

Experimental investigation of this process enables us to obtain information about the structure of that part of the lepton–hadron weak interaction Hamiltonian which does not conserve strangeness. We shall obtain the expression for the spectrum of π^0-mesons and leptons as well as for the polarization of the leptons.

The lepton–hadron interaction Hamiltonian in present V–A theory is given by the following expression [see (12.79)]:

$$\mathscr{H}_I(x) = \frac{G}{\sqrt{2}}\left[(\bar{\psi}_\mu(x)\,\gamma_\alpha(1+\gamma_5)\,\psi_{\nu_\mu}(x)) + (\bar{\psi}_e(x)\,\gamma_\alpha(1+\gamma_5)\,\psi_{\nu_e}(x))\right] j_\alpha(x) + \text{h.c.} \tag{18.2}$$

Here $\psi_\mu(x)$ $(\psi_e(x))$ is a muon (electron) field operator; $\psi_{\nu_\mu}(x)$ $(\psi_{\nu_e}(x))$ is a muon (electron) neutrino field operator, and $j_\alpha(x) = v_\alpha(x) + a_\alpha(x)$ is the hadron current. Denote the four-momentum of the initial K^--meson by q and the four-momenta of the π^0-meson, the charged lepton, and the antineutrino respectively by q', p, and p'. The final state vector can be written in the form

$$\Phi_{p,\,p',\,q'} = c_l^+(p)\,d_\nu^+(p')\Phi_{q'}, \tag{18.3}$$

where $\Phi_{q'}$ is the state vector of the π^0-meson. The state vector of the initial K^--meson we shall denote by Φ_q. In the first order of perturbation theory in the weak interaction constant G we have for the matrix element

$$\begin{aligned}(\Phi^+_{p,\,p',\,p'}S\Phi_q) = i\,\frac{G}{\sqrt{2}}\,\frac{1}{(2\pi)^3}\left(\frac{m_l}{p_0}\right)^{1/2}\bar{u}(p)\,\gamma_\alpha(1+\gamma_5)\,u(-p')\\ \times(\Phi^+_{q'}J_\alpha(0)\Phi_q)(2\pi)^4\,\delta(p'+q'+p-q).\end{aligned} \tag{18.4}$$

Here m_l is the mass of the charged lepton and $J_\alpha = V_\alpha + A_\alpha$ is the sum of the hadron vector and axial currents in the Heisenberg representation.

We shall ascertain the general structure of the hadron matrix element $(\Phi^+_{q'}J_\alpha(0)\Phi_q)$. We write this matrix element in the form

$$\begin{aligned}(\Phi^+_{q'}J_\alpha(0)\Phi_q) &= (\Phi^+_{q'}V_\alpha(0)\Phi_q) + (\Phi^+_{q'}A_\alpha(0)\Phi_q)\\ &= \frac{1}{(2\pi)^3}\,\frac{i}{\sqrt{4q_0q_0'}}\,(\Lambda^V_\alpha(q',q) + \Lambda^A_\alpha(q',q)).\end{aligned} \tag{18.5}$$

The quantities $\Lambda^V_\alpha(q',q)$ and $\Lambda^A_\alpha(q',q)$ transform like a vector and a pseudovector respectively. Since a pseudovector cannot be constructed from two four-vectors,

$$\Lambda^A_\alpha(q',q) = 0. \tag{18.6}$$

Thus, only the vector current V_α contributes to the matrix element of the process (18.1). It is obvious that the four-vector $\Lambda^V_\alpha(q',q)$ has the following general form:

$$\Lambda^V_\alpha(q',q) = f_+(q+q')_\alpha + f_-(q-q')_\alpha, \tag{18.7}$$

where f_+ and f_- are functions of the square of the four-momentum transfer $\varkappa^2 = (q-q')^2$. Thus

$$(\Phi^+_{q'}J_\alpha(0)\,\Phi_q) = \frac{1}{(2\pi)^3}\,\frac{i}{\sqrt{4q_0q_0'}}\,[f_+(\varkappa^2)(q+q')_\alpha + f_-(\varkappa^2)(q-q')_\alpha]. \tag{18.8}$$

The diagram of the process (18.1) with allowance for all strong interactions of the hadrons is shown in Fig. 24.

We now turn to the calculation of the decay probability. From (18.4) and (18.8) we obtain

$$d\Gamma = \frac{d\omega}{\varrho} = \frac{G^2}{2} 2\pi \left(\frac{m_l}{p_0}\right) \mathrm{Tr}\,[\gamma_\alpha(1+\gamma_5)\Lambda_\nu(-p')\overline{\gamma_\beta(1+\gamma_5)\Lambda(p)}]$$

$$\times(\Phi_{q'}^+ J_\alpha(0)\,\Phi_q)(\Phi_{q'}^+ J_\beta(0)\Phi_q)^*\,\delta(p'+p+q'-q)\,d\mathbf{p}\,d\mathbf{p}'\,d\mathbf{q}'. \quad (18.9)$$

Here

$$\Lambda(p) = \sum_s u^s(p)\,\bar{u}^s(p) = \frac{\hat{p}+im_l}{2im_l}, \quad \Lambda_\nu(-p') = \sum_r u^r(-p')\,\bar{u}^r(-p'). \quad (18.10)$$

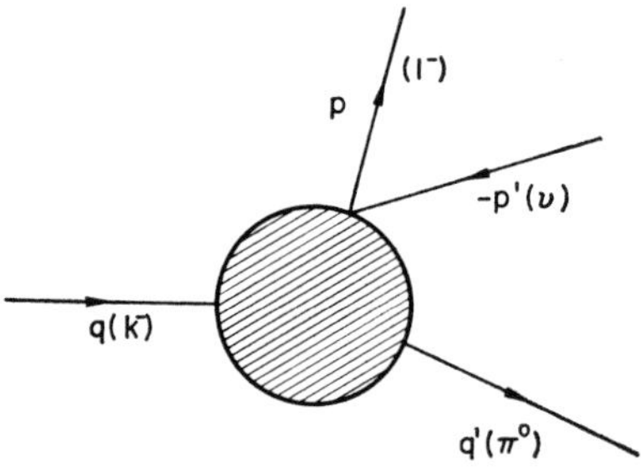

FIG. 24

Using the expression (obtained in the Appendix) for the density matrix of the antineutrino, we find

$$\Lambda_\nu(-p') = \tfrac{1}{2}(1+\gamma_5)\,\frac{\hat{p}'}{2ip_0'} + \tfrac{1}{2}(1-\gamma_5)\,\frac{\hat{p}'}{2ip_0'} = \frac{\hat{p}'}{2ip_0'}. \quad (18.11)$$

We first obtain the spectrum of π^0-mesons. For this we integrate over the momenta of the electron and the antineutrino. Considering that

$$\overline{\gamma_\beta(1+\gamma_5)} = \gamma_\beta(1+\gamma_5)\,\delta_\beta$$

($\delta_\beta = -1$ for $\beta = 1, 2, 3$ and $\delta_4 = 1$), we obtain

$$d\Gamma(q') = -\frac{\pi}{2}G^2\,\mathrm{Tr}\,[\gamma_\alpha(1+\gamma_5)\,\gamma_\varrho\gamma_\beta\gamma_\sigma]\,K_{\varrho\sigma}(\Phi_{q'}^+ J_\alpha(0)\Phi_q)\,(\Phi_{q'}^+ J_\beta(0)\,\Phi_q)^*\,\delta_\beta\,d\mathbf{q}', \quad (18.12)$$

where

$$K_{\varrho\sigma} = \int p_\varrho' p_\sigma\,\delta(p'+p-\varkappa)\,\frac{d\mathbf{p}'}{p_0'}\frac{d\mathbf{p}}{p_0}, \quad (18.13)$$

$$\varkappa = q-q'. \quad (18.14)$$

The integral (18.13) can be written out easily with the help of the methods discussed in § 17. It is obvious that $K_{\varrho\sigma}$ is a second-rank tensor which depends on the four-vector $\varkappa$. The tensor $K_{\varrho\sigma}$ has the general form

$$K_{\varrho\sigma} = a\delta_{\varrho\sigma}+b\varkappa_\varrho\varkappa_\sigma, \quad (18.15)$$

where a and b are functions of $\varkappa^2$. Multiplying (18.15) by $\delta_{\varrho\sigma}$ and summing over ϱ and σ, we find

$$\int (p'p)\,\delta(p'+p-\varkappa)\frac{d\mathbf{p}'}{p_0'}\frac{d\mathbf{p}}{p_0} = 4a+b\varkappa^2. \tag{18.16}$$

Using the law of four-momentum conservation we have

$$pp' = \frac{\varkappa^2+m_l^2}{2}. \tag{18.17}$$

From (18.16) and (18.17) we obtain

$$4a+b\varkappa^2 = \frac{\varkappa^2+m_l^2}{2}I, \tag{18.18}$$

where

$$I = \int \delta(p'+p-\varkappa)\frac{d\mathbf{p}'}{p_0'}\frac{d\mathbf{p}}{p_0}. \tag{18.19}$$

This integral was calculated in § 17 [see (17.25)]:

$$I = 2\pi\frac{\varkappa^2+m_l^2}{\varkappa^2}. \tag{18.20}$$

In order to obtain the second equation for a and b we multiply (18.15) by $\varkappa_\varrho\varkappa_\sigma$ and sum over ϱ and σ. Using the law of four-momentum conservation we find that

$$\varkappa p' = \frac{\varkappa^2+m_l^2}{2}, \qquad \varkappa p = \frac{\varkappa^2-m_l^2}{2}. \tag{18.21}$$

Thus, we obtain

$$a\varkappa^2+b(\varkappa^2)^2 = \frac{\varkappa^2-m_l^2}{2}\,\frac{\varkappa^2+m_l^2}{2}I. \tag{18.22}$$

From equations (18.18) and (18.22) we find that

$$a = \frac{\varkappa^2+m_l^2}{6\varkappa^2}K, \qquad b = \frac{\varkappa^2-2m_l^2}{3(\varkappa^2)^2}K. \tag{18.23}$$

Here

$$K = \frac{\varkappa^2+m_l^2}{2}I = \pi\frac{(\varkappa^2+m_l^2)^2}{\varkappa^2}. \tag{18.24}$$

We now calculate the trace in the expression for the decay probability. Using (18.15) we have

$$\mathrm{Tr}\,[\gamma_\alpha(1+\gamma_5)\,\gamma_\varrho\gamma_\beta\gamma_\sigma]\,K_{\varrho\sigma} = a\,\mathrm{Tr}\,[\gamma_\alpha(1+\gamma_5)\,\gamma_\varrho\gamma_\beta\gamma_\varrho]+b\,\mathrm{Tr}\,[\gamma_\alpha(1+\gamma_5)\,\hat{\varkappa}\gamma_\beta\hat{\varkappa}]. \tag{18.25}$$

Furthermore, with the help of the commutation relations for the γ matrices we obtain

$$\left.\begin{aligned} \gamma_\varrho\gamma_\beta\gamma_\varrho &= \gamma_\varrho(2\delta_{\beta\varrho}-\gamma_\varrho\gamma_\beta) = 2\gamma_\beta-4\gamma_\beta = -2\gamma_\beta, \\ \hat{\varkappa}\gamma_\beta\hat{\varkappa} &= \hat{\varkappa}(2\varkappa_\beta-\hat{\varkappa}\gamma_\beta) = 2\varkappa_\beta\hat{\varkappa}-\varkappa^2\gamma_\beta. \end{aligned}\right\} \tag{18.26}$$

Using these relations we find

$$\mathrm{Tr}\,[\gamma_\alpha(1+\gamma_5)\gamma_\varrho\gamma_\beta\gamma_\sigma]\,K_{\varrho\sigma} = 4[-(2a+b\varkappa^2)\delta_{\alpha\beta}+2b\varkappa_\alpha\varkappa_\beta]. \tag{18.27}$$

Let us write the matrix element (18.8) in the form

$$(\Phi^+_{q'}J_\alpha(0)\Phi_q) = i\frac{1}{(2\pi)^3}\frac{1}{\sqrt{4q_0q_0'}}f_+[2q_\alpha+(\eta-1)\,\varkappa_\alpha], \tag{18.28}$$

where

$$\eta = \frac{f_-}{f_+}. \tag{18.29}$$

Noting that

$$q^*_\beta\,\delta_\beta = -q_\beta \tag{18.30}$$

(there is no summation over β here), from (18.12) and (18.27) we obtain

$$d\Gamma(q') = \frac{G^2}{(2\pi)^5}\frac{1}{\sqrt{4q_0q_0'}}\,|f_+|^2\{4(2a+b\varkappa^2)\,m_K^2 + 8b(q\varkappa)^2+(-2a+b\varkappa)^2\,[4q\varkappa\,\mathrm{Re}\,(\eta-1)+\varkappa^2|\eta-1\,|^2]\}\,dq'. \tag{18.31}$$

The final expression for the π^0-meson spectrum is written in the rest system of the initial K-mesons. Integrating over the emission angles of the π^0-mesons, from (18.31), (18.23), and (18.24) we obtain

$$\frac{d\Gamma}{dE_\pi} = \frac{G^2}{(2\pi)^3}\frac{1}{m_K}\,|f_+|^2\,(E_\pi^2-m_\pi^2)^{1/2}\left(\frac{\varkappa^2+m_l^2}{\varkappa^2}\right)^2\left\{m_K^2\varkappa^2-\tfrac{1}{3}m_K^2(\varkappa^2+m_l^2) +\tfrac{2}{3}\left(1-\frac{2m_l^2}{\varkappa^2}\right)(q\varkappa)^2-m_l^2[q\varkappa\,\mathrm{Re}\,(\eta-1)+\varkappa^2\tfrac{1}{4}\,|\eta-1\,|^2]\right\}, \tag{18.32}$$

where E_π is the total energy of the π-mesons in the K-meson rest system. Obviously

$$\left.\begin{aligned}\varkappa q &= \tfrac{1}{2}(-m_K^2+m_\pi^2+\varkappa^2) = -m_K(m_K-E_\pi),\\ \varkappa^2 &= -m_K^2-m_\pi^2+2m_KE_\pi\end{aligned}\right\} \tag{18.33}$$

(m_π and m_K are the masses of the π- and K-mesons).

Let us find the limits within which the quantities $\varkappa^2$ and E_π vary. Since $E_\pi \geqslant m_\pi$,

$$\varkappa^2 \geqslant -(m_K-m_\pi)^2. \tag{18.34}$$

To find the upper limit of the variable $\varkappa^2$ we note that

$$\varkappa^2 = (p'+p)^2 = -\left(p_0'+\sqrt{m_l^2+(p_0')^2}\right)^2,$$

where p_0' is the energy of the neutrino in the c.m. system of the lepton and the neutron. Since $p_0' \geqslant 0$, from this we obtain

$$\varkappa^2 \leqslant -m_l^2. \tag{18.35}$$

From (18.33) and (18.35) we find

$$E_\pi \leqslant \frac{m_K^2+m_\pi^2-m_l^2}{2m_K}. \tag{18.36}$$

Note that

$$\varkappa^2+m_l^2 = -2m_K(E_\pi^0-E_\pi), \tag{18.37}$$

where

$$E_\pi^0 = \frac{m_K^2+m_\pi^2-m_l^2}{2m_K} \tag{18.38}$$

is the maximum energy of the π-meson created in the decay $K \to \pi+l+\nu$. The factor $(E_\pi^2-m_\pi^2)^{1/2}(\varkappa^2+m_l^2)^2$ in expression (18.32) vanishes consequently on the boundaries of the π-meson spectrum.

In conclusion we make the following observation. If, using the conservation of four-momentum, we replace $\varkappa$ by the vector $p+p'$ and use the Dirac equation for the spinors $\bar{u}(p)$ and $u(-p')$, it is easy to see that

$$f_-\bar{u}(p)\,\gamma_\alpha(1+\gamma_5)\,u(-p')\,\varkappa_\alpha = im_l f_-\bar{u}(p)(1+\gamma_5)\,u(-p').$$

Thus, the form factor f_- enters the matrix element of the process under discussion in the form $m_l f_-$. In the expression for the π-meson spectrum the form factor f_-, as can be seen from (18.32), is multiplied by the square of the lepton mass. We shall give an expression for that part of the π-meson spectrum from the decay $K^- \to \pi^0+e^-+\bar{\nu}_e$ for which $|\varkappa|^2 \gg m_l^2$. Omitting terms of order $(m_l/m_K)^2$ and $m_l^2/\varkappa^2$, we find from (18.32)

$$\left(\frac{d\Gamma}{dE_\pi}\right)_{K\to\pi e\nu} = \frac{G^2}{12\pi^3}\,|f_+|^2\, m_K(E_\pi^2-m_\pi^2)^{3/2}. \tag{18.39}$$

Experimental study of the π^0-meson spectra from the decays $K\to\pi+\mu+\nu_\mu$ and $K\to\pi+e+\nu_e$ enables us to investigate the dependence of the form factors f_+ and f_- on the square of the momentum transfer $\varkappa^2$. These form factors are usually represented in the form

$$f_\pm(\varkappa^2) = f_\pm(0)\left(1+\lambda_\pm\frac{\varkappa^2}{m_\pi^2}\right), \tag{18.40}$$

where $\lambda_\pm$ are dimensionless parameters characterizing the dependence of the form factors on $\varkappa^2$. From the data of various experiments it is known that $\lambda_+ \leqslant 0.004$.

Let us now obtain the expressions for the spectrum and polarization of the leptons created in the decays (18.1). For this we disregard the dependence of the form factors f_+ and f_- on the square of the four-momentum transfer. Using the Dirac equation for the spinors $\bar{u}(p)$ and $u(-p')$ we write the matrix element of the process in the form

$$(\Phi^+_{p,\,p',\,q'}S\Phi_q) = \frac{1}{(2\pi)^2}\left(\frac{m_l}{4p_0q_0q_0'}\right)^{1/2}\bar{u}(p)\,\mathfrak{M}u(-p')\,\delta(p'+q'-k). \tag{18.41}$$

Here

$$\mathfrak{M} = \frac{G}{\sqrt{2}}f_+(2\hat{k}+im_l(\eta+1))(1+\gamma_5), \tag{18.42}$$

$$k = q-p. \tag{18.43}$$

The mean value of the spin operator of a charged lepton under the condition that no π^0-mesons and neutrinos are recorded is equal to

$$\xi_\mu = \frac{N_\mu}{R}, \tag{18.44}$$

where

$$N_\mu = \int \frac{m_l}{q_0'} \sum_{s, s', r} (\bar{u}^{s'}(p)\, i\gamma_5\gamma_\mu u^s(p))(\bar{u}^s(p)\, \mathfrak{M} u^r(-p'))(\bar{u}^r(-p')\, \overline{\mathfrak{M}} u^{s'}(p))\, \delta(p'+q'-k)\, d\boldsymbol{p}'\, d\boldsymbol{q}', \tag{18.45}$$

$$R = \int \frac{m_l}{q_0'} \sum_{s, r} (\bar{u}^s(p)\, \mathfrak{M} u^r(-p'))(\bar{u}^r(-p')\, \overline{\mathfrak{M}} u^s(p))\, \delta(p'+q'-k)\, d\boldsymbol{p}'\, d\boldsymbol{q}'. \tag{18.46}$$

The lepton spectrum is given by an expression proportional to R. Let us obtain this spectrum. Using (18.10) we find that the probability of emitting a lepton with momentum in the interval from $\boldsymbol{p}$ to $\boldsymbol{p}+d\boldsymbol{p}$ is equal to

$$d\Gamma(p) = \frac{1}{(2\pi)^5} \frac{1}{4q_0 p_0} R\, d\boldsymbol{p} = \frac{1}{(2\pi)^5} \frac{1}{4q_0} \times \left[\int \frac{m_l}{q_0'} \mathrm{Tr}\, [\mathfrak{M} \Lambda_\nu(-p')\, \overline{\mathfrak{M}} \Lambda(p)]\, \delta(p'+q'-k)\, d\boldsymbol{p}'\, d\boldsymbol{q}' \right] \frac{d\boldsymbol{p}}{p_0}. \tag{18.47}$$

We integrate over the momenta of the antineutrino and the π^0-meson. Since the matrix $\mathfrak{M}$ does not depend on the integration variables, we have

$$R = \frac{m_l}{2i} \mathrm{Tr}\, [\mathfrak{M} I_\alpha \gamma_\alpha \overline{\mathfrak{M}} \Lambda(p)]. \tag{18.48}$$

Here

$$I_\alpha = \int p_\alpha' \, \delta(p'+q'-k) \frac{d\boldsymbol{p}'}{p_0'} \frac{d\boldsymbol{q}'}{q_0'}. \tag{18.49}$$

This integral was calculated in the preceding section [see (17.28)]. We have

$$I_\alpha = I_0 k_\alpha, \tag{18.50}$$

where

$$I_0 = \pi \left(\frac{k^2 + m_\pi^2}{k^2} \right)^2. \tag{18.51}$$

It is, furthermore, easy to see that

$$\overline{\mathfrak{M}} = -\frac{G}{\sqrt{2}} f_+^*(1-\gamma_5)\, [2\hat{k} + im_l(\eta^* + 1)]. \tag{18.52}$$

With the help of (18.42), (18.50), and (18.52) we find

$$R = \tfrac{1}{4} G^2 |f_+|^2 I_0\, \mathrm{Tr} \left\{ (2\hat{k} + im_l(\eta+1))(1+\gamma_5)[2k^2 + im_l\hat{k}(\eta^*+1)]\, (\hat{p} + im_l) \right\}. \tag{18.53}$$

The calculation of this trace does not present any difficulty. We obtain

$$R = G^2|f_+|^2 I_0[4k^2(kp)-4k^2m_l^2 \operatorname{Re}(\eta+1)-m_l^2|\eta+1|^2 kp]. \tag{18.54}$$

From (18.47), (18.48), (18.51), and (18.54) we find the following expression for the lepton energy spectrum in the K-meson rest system:

$$\frac{d\Gamma}{dE} = \frac{1}{(2\pi)^3}\frac{1}{m_K} G^2|f_+|^2(E^2-m_l^2)^{1/2}\frac{(k^2+m_\pi^2)^2}{(-k^2)}$$
$$\times\left[m_K E+m_l^2 \operatorname{Re}\eta-\frac{m_l^2|\eta+1|^2(m_K E-m_l^2)}{4k^2}\right]. \tag{18.55}$$

Here E is the total energy of the lepton in the K-meson rest system and

$$k^2 = -m_K^2-m_l^2+2m_K E. \tag{18.56}$$

Let us find the limits on the variable k^2. Using the conservation of four-momentum in the c.m. system of the neutrino and π-meson, we obtain

$$k^2 = (q'+p')^2 = -\left(p_0'+\sqrt{m_\pi^2+(p_0')^2}\right). \tag{18.57}$$

Hence

$$k^2 \leqslant -m_\pi^2. \tag{18.58}$$

On the other hand, since $E \geqslant m_1$, we find from (18.56)

$$k^2 \geqslant -(m_K-m_l)^2. \tag{18.59}$$

With the help of (18.56) and (18.58) we find that the maximum energy of the lepton in the K-meson rest system is equal to

$$E_0 = \frac{m_K^2+m_l^2-m_\pi^2}{2m_K}. \tag{18.60}$$

Note that

$$k^2+m_\pi^2 = -2m_K(E_0-E). \tag{18.61}$$

In the case of the decay $K^-\to\pi^0+e^-+\nu_e$ one can disregard terms proportional to m_e^2 in the square brackets of expression (18.55). The electron spectrum from the decay $K^-\to\pi^0+e^-+\nu_e$ does not depend therefore on the parameter η and is given by the expression

$$\left(\frac{d\Gamma}{dE}\right)_{K\to\pi e\nu} = \frac{1}{4\pi^3}G^2|f_+|^2 m_K E(E^2-m_e^2)^{1/2}\frac{(E_0-E)^2}{\left[(E_0-E)+\dfrac{m_\pi^2}{2m_K}\right]}. \tag{18.62}$$

We now calculate the μ-meson (electron) polarization vector. Using (18.10) and (18.50) we easily find that the numerator of expression (18.44) is equal to

$$N_\mu = \int\frac{m_l}{q_0'}\operatorname{Tr}[i\gamma_5\gamma_\mu\Lambda(p)\,\mathfrak{M}\Lambda_\nu(-p')\,\overline{\mathfrak{M}}\Lambda(p)]$$
$$\times\delta(p'+q'-k)\,dp'\,dq' = \frac{1}{2i}m_l I_0\operatorname{Tr}[i\gamma_5\gamma_\mu\Lambda(p)\,\mathfrak{M}\hat{k}\overline{\mathfrak{M}}\Lambda(p)]. \tag{18.63}$$

Obviously

$$N_\mu p_\mu = 0. \tag{18.64}$$

The calculation of the trace in (18.63) is simplified if we use the relationship

$$\gamma_5\gamma_\mu\Lambda(p) = \Lambda(p)\,\gamma_5\gamma_\mu + \frac{p_\mu}{im_l}\gamma_5, \tag{18.65}$$

which can be obtained easily from the commutation relations for the γ matrices. Considering that

$$\Lambda(p)\,\Lambda(p) = \Lambda(p),$$

we find

$$N_\mu = \frac{1}{2i}m_l I_0\left\{\mathrm{Tr}\,[i\gamma_5\gamma_\mu\mathfrak{M}\hat{k}\overline{\mathfrak{M}}\Lambda(p)] + \frac{p_\mu}{m_l}\mathrm{Tr}\,[\gamma_5\mathfrak{M}\hat{k}\overline{\mathfrak{M}}\Lambda(p)]\right\}. \tag{18.66}$$

The first trace is equal to

$$\frac{1}{2i}m_l\,\mathrm{Tr}\,[i\gamma_5\gamma_\mu\mathfrak{M}\hat{k}\overline{\mathfrak{M}}\Lambda(p)] = -|f_+|^2G^2m_l\,[(4k^2+m_l^2|\eta+1|^2)\,k_\mu - 4ik^2\,\mathrm{Im}\,\eta p_\mu]. \tag{18.67}$$

There is no need to calculate the second trace. Using (18.64) we obtain

$$\mathrm{Tr}\,[\gamma_5\mathfrak{M}\hat{k}\overline{\mathfrak{M}}\Lambda(p)] = \frac{1}{m_l}p_\mu\,\mathrm{Tr}\,[i\gamma_5\gamma_\mu\mathfrak{M}\hat{k}\overline{\mathfrak{M}}\Lambda(p)]. \tag{18.68}$$

With the help of (18.66) and (18.68) we find

$$N_\mu = -|f_+|^2G^2I_0\,(4k^2+m_l^2|\eta+1|^2)\,m_l\left(k_\mu + p_\mu\frac{kp}{m_l^2}\right). \tag{18.69}$$

For the lepton polarization four-vector ξ_μ we obtain from (18.68) and (18.54) the expression

$$\xi_\mu = -\frac{(4k^2+m_l^2|\eta+1|^2)\,m_l\left(k_\mu + p_\mu\dfrac{kp}{m_l^2}\right)}{[(4k^2-m_l^2|\eta+1|^2)\,kp - 4k^2m_l^2\,\mathrm{Re}\,(\eta+1)]}. \tag{18.70}$$

Let us calculate the polarization vector in the lepton rest system. Choosing the x-axis along the lepton momentum $\boldsymbol{p}$, we have

$$\xi_x^0 = \frac{\xi_x - \beta\xi_0}{\sqrt{1-\beta^2}}, \quad \xi_y^0 = \xi_y, \quad \xi_z^0 = \xi_z. \tag{18.71}$$

Here $\boldsymbol{\xi}^0$ is the polarization vector in the lepton rest system and $\beta = |\boldsymbol{p}|/p_0$. From the condition $\xi p = 0$ it is obvious that $\xi_0^0 = 0$. It also follows from this condition that

$$\xi_0 = \beta\xi_x. \tag{18.72}$$

With the help of (18.71) and (18.72) we find that the polarization vector in the lepton rest system is related to the particle's polarization vector in the system where the momentum of the lepton is $\boldsymbol{p}$ by

$$\boldsymbol{\xi}^0 = \boldsymbol{\xi} - \frac{1}{p_0(p_0+m_l)}(\boldsymbol{\xi p})\boldsymbol{p}. \tag{18.73}$$

Furthermore, it is not difficult to see that

$$k_\mu + p_\mu \frac{kp}{m_l^2} = q_\mu + p_\mu \frac{qp}{m_l^2}. \tag{18.74}$$

The space part of this vector in the rest system of the initial K-meson is equal to

$$-\frac{1}{m_l^2} m_K E \boldsymbol{p}.$$

After these last remarks we can easily find from (18.70) and (18.73) that the lepton polarization vector in the rest system which is obtained from the rest system of the initial K-meson by a Lorentz transformation along the momentum is equal to

$$\boldsymbol{\xi}^0 = -\frac{\left(1 + \frac{m_l^2|\eta+1|^2}{4k^2}\right) m_K \boldsymbol{p}}{\left[m_K E + m_l^2 \operatorname{Re}\eta - \frac{m_l^2|\eta+1|^2}{4k^2}(m_K E - m_l^2)\right]}. \tag{18.75}$$

Here $\boldsymbol{p}$ and E are the momentum and energy of the lepton in the K-meson rest system and k^2 is given by expression (18.56).

In the case of decays $K^- \to \pi^0 + e^- + \bar{\nu}_e$ one can disregard terms proportional to m_e^2 and then obtain for the electron polarization

$$\boldsymbol{\xi}^0 = -\frac{\boldsymbol{p}}{E}. \tag{18.76}$$

Thus, the electron polarization in the decay $K^- \to \pi^0 + e^- + \bar{\nu}_e$ does not depend on the parameter η and is equal to $-\boldsymbol{v}$. Measurement of the polarization and spectrum of μ-mesons makes it possible to obtain information about the size of the parameter η.

In conclusion we make the following remark. Up to this point we have supposed that the form factors f_+ and f_- were complex. It is not difficult to show that if there is invariance under time-reversal, the form factors are real. Actually, from the condition of unitarity of the S-matrix (in the lowest order of the weak interaction constant G) and from T-invariance, analogous to (12.105) we obtain

$$(\Phi^+_{q'_T} J_\alpha(0)\, \Phi_{q_T})^* = (\Phi^+_{q'} J_\alpha(0)\, \Phi_q). \tag{18.77}$$

Here $q_T = (-\boldsymbol{q}, iq_0)$; $q'_T = (-\boldsymbol{q}', iq'_0)$.

Using (18.5), (18.7) and (18.77) we find that

$$f_+^* = f_+, \quad f_-^* = f_-. \tag{18.78}$$

19. The Decay $\pi^0 \to 2\gamma$

In conclusion we examine the decay of a neutral π-meson into two photons:

$$\pi^0 \to \gamma+\gamma. \tag{19.1}$$

Denote the initial and final state vectors by $\Phi_{k_1\lambda_1,\, k_2\lambda_2}$ and Φ_q (k_1 and k_2 are momenta and λ_1 and λ_2 are the indices of the polarization vectors of the final photons; q is the momentum of the initial π^0-meson). Obviously

$$\Phi_{k_1\lambda_1,\, k_2\lambda_2} = a^+_{\lambda_1}(k_1)\, a^+_{\lambda_2}(k_2)\, \Phi_0. \tag{19.2}$$

Both electromagnetic and strong interactions are responsible for the decay (19.1). Strong interactions cannot be examined within perturbation theory. We shall construct a general expression for the matrix element of the process and then calculate the decay probability.

The diagram for the decay process (19.1) taking account of all interactions responsible for the decay is shown in Fig. 25.

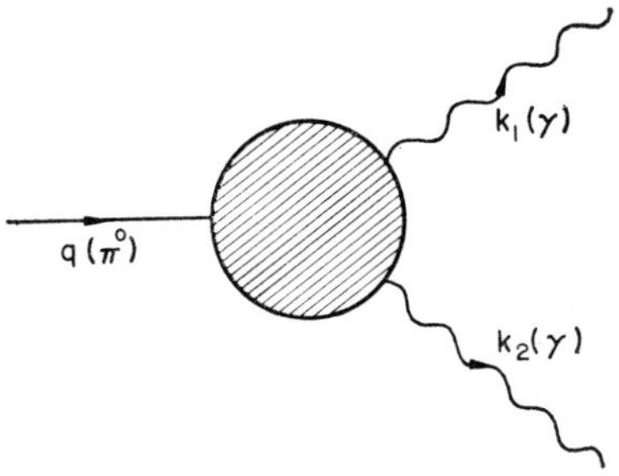

FIG. 25

The matrix element corresponding to this diagram can be written

$$(\Phi^+_{k_1\lambda_1,\, k_2\lambda_2} S\Phi_q) = e^2 \frac{1}{(2\pi)^{9/2}} \frac{1}{\sqrt{8k_{10}k_{20}q_0}} e^{\lambda_1}_\mu(k_1)$$
$$\times e^{\lambda_2}_\nu(k_2) \mathfrak{M}_{\mu\nu}(k_1,\, k_2;\, q)(2\pi)^4\, \delta(k_1+k_2-q), \tag{19.3}$$

where $e^{\lambda_1}(k_1)$ and $e^{\lambda_2}(k_2)$ are the polarization four-vectors of the final photons. In expression (19.3) the standard factors corresponding to the external photon and meson lines are isolated; the quantity $\mathfrak{M}_{\mu\nu}(k_1,\, k_2;\, q)$ corresponds to the hatched part of the diagram in Fig. 25 and is determined by strong interactions. The factor e^2 is also isolated in (19.3).

Let us construct the general expression for $\mathfrak{M}_{\mu\nu}(k_1,\, k_2;\, q)$. For this we shall ascertain which conditions this quantity must satisfy. First of all as a result of the commutation of the operators $a_{\lambda_1}(k_1)$ and $a_{\lambda_2}(k_2)$ we have

$$(\Phi^+_{k_1\lambda_1,\, k_2\lambda_2} S\Phi_q) = (\Phi^+_{k_2\lambda_2,\, k_1\lambda_1} S\Phi_q). \tag{19.4}$$

From (19.3) and (19.4) we conclude that the indistinguishability of the photons leads to the relationship

$$\mathfrak{M}_{\mu\nu}(k_1,\, k_2;\, q) = -\mathfrak{M}_{\nu\mu}(k_2,\, k_1;\, q). \tag{19.5}$$

Furthermore, from requirements of Lorentz invariance and invariance under inversion it follows that the quantity $\mathfrak{M}_{\mu\nu}(k_1, k_2; q)$ must transform like a second-rank pseudotensor (π^0 is a pseudoscalar particle).

Let us construct from the momentum vectors a quantity satisfying these conditions. Obviously as a result of the laws of energy-momentum conservation

$$q = k_1 + k_2 \tag{19.6}$$

only two momentum four-vectors are independent. We shall choose as independent vectors k_1 and k_2. From two four-vectors only one second-rank pseudotensor can be constructed. Thus, in the most general case

$$\mathfrak{M}_{\mu\nu}(k_1, k_2; q) = a\varepsilon_{\mu\nu\varrho\sigma}k_{1\varrho}k_{2\sigma}, \tag{19.7}$$

where the quantity a depends on the scalars which can be constructed from k_1 and k_2. Since

$$k_1^2 = k_2^2 = 0, \quad k_1k_2 = \tfrac{1}{2}q^2 = -\tfrac{1}{2}m_\pi^2 \tag{19.8}$$

(m_π is the mass of the π-meson), a is a constant. It is obvious that (19.7) satisfies relationship (19.5).

We now turn to an examination of the conditions imposed on the matrix element of the process by the law of current conservation. First we discuss the general case. Let us examine some process in which a photon is created. Denote by Φ_i the state vector of the initial particles. We write the final state vector

$$a_\lambda^+(k)\,\Phi_f, \tag{19.9}$$

where k and λ are the four-momentum and index of the polarization vector of the photon ($\lambda = 1, 2$); Φ_f is the state vector of the remaining final particles. The matrix element of the process has the form

$$R_{k\lambda, f; i} = (\Phi_f^+ a_\lambda(k)\,(S-1)\,\Phi_i). \tag{19.10}$$

Obviously

$$a_\lambda(k)\,(S-1) = [a_\lambda(k), S] + (S-1)\,a_\lambda(k). \tag{19.11}$$

We assume that there are no photons with momentum k in the initial state. The second term in (19.11) does not contribute to the matrix element of the process. Let us examine the first term. Using expression (8.31) for the S-matrix we have

$$[a_\lambda(k), S] = \sum_{n=1}^{\infty} \frac{(-i)^n}{n!} \int [a_\lambda(k), T(\mathscr{H}_I(x_1)\ldots\mathscr{H}_I(x_n))]\, dx_1\ldots dx_n. \tag{19.12}$$

With the help of (4.28) it is not difficult to show that for any operators $\alpha, \beta, \gamma, \ldots, \delta$ there exists the relationship

$$[\alpha, \beta\gamma\ldots\delta] = [\alpha, \beta]\,\gamma\ldots\delta + \beta[\alpha, \gamma]\ldots\delta + \ldots + \beta\gamma\ldots[\alpha, \delta]. \tag{19.13}$$

Using this relationship, permuting the operators under the T-product symbol, and reassigning the integration variables, we obtain

$$[a_\lambda(k), S] = -i \int T([a_\lambda(k), \mathcal{H}_I(x)]\, S)\, dx. \tag{19.14}$$

It is obvious that only the electromagnetic interaction Hamiltonian makes a non-zero contribution to the commutator $[a_\lambda(k), \mathcal{H}_I(x)]$.

Let us take as an example the interaction of an electromagnetic field with proton and π-meson fields. The interaction Hamiltonian is given by expression (12.3). We obtain

$$[a_\lambda(k), \mathcal{H}_I(x)] = -[((-e)j^p_\mu(x)+ej^\pi_\mu(x))]\, [a_\lambda(k), A_\mu(x)]. \tag{19.15}$$

The currents $j^p_\mu(x)$ and $j^\pi_\mu(x)$ are given by expressions (12.4); e is the charge of the electron. In the general case

$$[a_\lambda(k), \mathcal{H}_I(x)] = -ej_\mu(x)\, [a_\lambda(k), A_\mu(x)], \tag{19.16}$$

where $j_\mu(x)$ is the current of all charged particles. With the help of (7.35) and (7.40) we find easily

$$[a_\lambda(k), A_\mu(x)] = \frac{1}{(2\pi)^{3/2}} \frac{1}{\sqrt{2k_0}}\, e^\lambda_\mu(k)\, \mathrm{e}^{-ikx}. \tag{19.17}$$

From (19.11), (19.14), (19.16), and (19.17) we find that the matrix element of the process is

$$R_{k\lambda, f; i} = ie \frac{1}{(2\pi)^{3/2}} \frac{1}{\sqrt{2k_0}}\, e^\lambda_\mu(k) \int \mathrm{e}^{-ikx} \big(\Phi^+_f T(j_\mu(x)\, S)\, \Phi_i\big)\, dx. \tag{19.18}$$

Transforming to the Heisenberg representation (see § 12) we have

$$\big(\Phi^+_f T(j_\mu(x)\, S)\, \Phi_i\big) = (\Phi^+_{f;\,\mathrm{out}} J_\mu(x)\, \Phi_{i;\,\mathrm{in}}), \tag{19.19}$$

where $J_\mu(x)$ is the current in the Heisenberg representation.

Furthermore, analogous to (12.32) we find

$$(\Phi^+_{f;\,\mathrm{out}} J_\mu(x)\, \Phi_{i;\,\mathrm{in}}) = \mathrm{e}^{-i(P_f-P_i)x}\, (\Phi^+_{f;\,\mathrm{out}} J_\mu(0)\, \Phi_{i;\,\mathrm{in}}), \tag{19.20}$$

where P_f and P_i are the total four-momenta of the particles described respectively by the vectors Φ_f and Φ_i. Inserting (19.19) and (19.20) into expression (19.18) and integrating over x, we obtain for the matrix element of the process

$$R_{k\lambda, f; i} = ie \frac{1}{(2\pi)^{3/2}} \frac{1}{\sqrt{2k_0}}\, e^\lambda_\mu(k)\, (\Phi^+_{f;\,\mathrm{out}} J_\mu(0) \Phi_{i;\,\mathrm{in}})\, (2\pi)^4\, \delta(P_f+k-P_i). \tag{19.21}$$

With the help of (19.20) it is not difficult to see that from the law of current conservation

$$\frac{\partial J_\mu(x)}{\partial x_\mu} = 0 \tag{19.22}$$

it follows that

$$(P_f-P_i)_\mu (\Phi^+_{f;\,\mathrm{out}} J_\mu(0)\, \Phi_{i;\,\mathrm{in}}) = 0. \tag{19.23}$$

We write the matrix element for the process in the form

$$R_{k\lambda,\, f;\, i} = e_\mu^\lambda(k) M_\mu(k, f;\, i)\, \delta(P_f + k - P_i). \tag{19.24}$$

From (19.21), (19.23), and (19.24) we conclude that the quantity $M_\mu(k, f;\, i)$ must satisfy the following condition:

$$k_\mu M_\mu(k, f;\, i) = 0. \tag{19.25}$$

In the case of the decay $\pi^0 \to 2\gamma$ the relationships

$$\left.\begin{aligned} k_{1\mu}\mathfrak{M}_{\mu\nu}(k_1, k_2;\, q) &= 0, \\ k_{2\mu}\mathfrak{M}_{\mu\nu}(k_1, k_2;\, q) &= 0 \end{aligned}\right\} \tag{19.26}$$

must therefore be satisfied.

It is easy to see that expression (19.7) which was constructed from the requirements of Lorentz invariance and invariance under inversion automatically satisfies (19.26).

We now find the decay probability of the neutral π-meson into two photons. To calculate the total decay probability it is necessary to sum over the indices λ_1 and λ_2. From (19.3) we find in the π^0-meson rest system the following expression for the decay probability:

$$\begin{aligned} d\Gamma = {} & \frac{1}{(2\pi)^2} \frac{1}{8m_\pi} e^4 \sum_{\lambda_1,\, \lambda_2 = 1, 2} \left(e_\mu^{\lambda_1}(k_1)\, e_\nu^{\lambda_2}(k_2)\, \mathfrak{M}_{\mu\nu}\right) \\ & \times \left(e_\varrho^{\lambda_1}(k_1)\, e_\sigma^{\lambda_2}(k_2)\, \mathfrak{M}_{\varrho\sigma}\right)^* \delta(k_1 + k_2 - q) \frac{d\boldsymbol{k}_1}{k_{10}} \cdot \frac{d\boldsymbol{k}_2}{k_{20}}. \end{aligned} \tag{19.27}$$

We sum over the polarization states of the photons. Real photons emitted in the process (19.1) can possess only transverse polarization. As a result of this the summation in (19.27) is only over λ_1 and λ_2 equal to 1 and 2. We shall show that the condition (19.25) which follows from the law of current conservation enables us to extend the summation to all values of λ.

Let us first discuss the general case. We examine some process which produces photons. The matrix element of the process has the form (19.24). In order to obtain the probability of the process summed over the polarization states of the photon it is necessary to calculate the sum

$$\sum_{\lambda = 1, 2} (e^\lambda(k)\, M)(e^\lambda(k)\, M)^*. \tag{19.28}$$

The vectors $e^1(k)$ and $e^2(k)$ which describe the states of transverse polarization of the photon are defined in the following manner [see (7.13)]:

$$e^\lambda(k) = (\boldsymbol{e}^\lambda(\boldsymbol{k}), 0) \qquad (\lambda = 1, 2), \tag{19.29}$$

where

$$\boldsymbol{e}^\lambda(\boldsymbol{k})\, \boldsymbol{k} = 0.$$

The vectors of longitudinal and scalar polarization are

$$e_\mu^3(k) = \left(\frac{\boldsymbol{k}}{k_0}, 0\right), \quad e^4(k) = (0, i). \tag{19.30}$$

Obviously

$$e_\mu^3(k)+e_\mu^4(k) = \frac{k_\mu}{k_0}. \tag{19.31}$$

From (19.31) and (19.25) we obtain

$$(e^3(k)\,M) = -(e^4(k)\,M). \tag{19.32}$$

Hence

$$(e^3(k)\,M)(e^3(k)\,M)^* = (e^4(k)\,M)(e^4(k)\,M)^*. \tag{19.33}$$

Thus, condition (19.25) implies that the probabilities of emission of longitudinal and scalar photons are equal. With the help of (19.33) we obtain

$$\sum_{\lambda=1,2} (e^\lambda(k)\,M)(e^\lambda(k)\,M)^* = \sum_{\lambda=1}^{4} (e^\lambda(k)\,M)(e^\lambda(k)\,M)^*\,\eta_\lambda. \tag{19.34}$$

(The factor η_λ is equal to $+1$ for $\lambda = 1, 2, 3$, and -1 for $\lambda = 4$.)

Now we can use the condition of completeness of the set of vectors $e^\lambda(k)$:

$$\sum_{\lambda=1}^{4} e_\mu^\lambda(k)\,e_\nu^\lambda(k)\,\eta_\lambda = \delta_{\mu\nu}. \tag{19.35}$$

Noting that

$$(e_\nu^\lambda(k))^* = e_\nu^\lambda(k)\,\eta_\nu,$$

we find from (19.34) and (19.35) without difficulty

$$\sum_{\lambda=1,2} (e^\lambda(k)\,M)(e^\lambda(k)\,M)^* = \sum_\mu M_\mu M_\mu^*\eta_\mu. \tag{19.36}$$

This final result can be presented in another form. We write the sum over λ in (19.36) in the following manner:

$$\sum_{\mu,\,\nu} M_\mu M_\nu^*\eta_\nu \sum_{\lambda=1,2} e_\mu^\lambda(k)\,e_\nu^\lambda(k). \tag{19.37}$$

Obviously relationship (19.26) can be obtained if we set

$$\sum_{\lambda=1,2} e_\mu^\lambda(k)\,e_\nu^\lambda(k) = \delta_{\mu\nu} \tag{19.38}$$

(this is a conditional equality; it is to be understood in the following manner: if we contract the sum $\sum_{\lambda=1,2} e_\mu^\lambda(k)\,e_\nu^\lambda(k)$ with the tensor $M_\mu M_\nu^*\eta_\nu$, then as a result of (19.25) the sum can be replaced by $\delta_{\mu\nu}$).

We make the following comment in connection with what has been said. The vectors (19.29) and (19.30) are defined in a fixed reference system. It is not difficult to introduce covariantly defined four-vectors describing the polarization states of the photon. Denote by n a time-like unit four-vector:

$$n^2 = -1. \tag{19.39}$$

Let

$$e^4 = n. \tag{19.40}$$

The vectors describing transverse polarization were defined so that

$$e^\lambda(k)\, n = 0, \tag{19.41}$$

$$e^\lambda(k)\, k = 0 \qquad (\lambda = 1,\ 2). \tag{19.42}$$

[Condition (19.41) means that the $e^\lambda(k)$ where $\lambda = 1, 2$ are space-like vectors; $e_4^\lambda(k) = 0$ in the system where $n = (0, i)$; from condition (19.42) it follows that in this system $e^\lambda(k)\boldsymbol{k} = 0$.] The longitudinal polarization vector $e^3(k)$ must be orthogonal to all remaining constructed vectors. Assume

$$e_\mu^3(k) = ck_\mu + dn_\mu. \tag{19.43}$$

From (19.41) and (19.42) it follows that this vector is orthogonal to the vectors $e^1(k)$ and $e^2(k)$. Furthermore, requiring the orthogonality of $e^3(k)$ and n we obtain

$$d = c(kn). \tag{19.44}$$

Finally, from the condition $e^3e^3 = 1$ we find

$$c^2 = \frac{1}{(kn)^2}. \tag{19.45}$$

Suppose that

$$e^3(k) = -\left(n + \frac{k}{kn}\right). \tag{19.46}$$

Obviously the four-vectors constructed in the reference system where $n = (0, i)$ coincide with the vectors (19.29) and (19.30). From (19.40) and (19.46) we find that

$$e^3(k) + e^4(k) = \frac{-k}{kn}. \tag{19.47}$$

Obviously with the help of (19.25) and (19.47) we obtain equations (19.32) and (19.33) and furthermore, using the completeness of the constructed set of vectors, we arrive at (19.36).

Let us now return to the calculation of the decay probability of the neutral π-meson into two photons. From (19.27) with the help of (19.36) we find

$$d\Gamma = \frac{1}{(2\pi)^2}\,\frac{1}{8m_\pi}\, e^4 \sum_{\mu,\, \nu} \mathfrak{M}_{\mu\nu}\mathfrak{M}_{\mu\nu}^{*}\eta_\mu\eta_\nu\, \delta(k_1 + k_2 - q)\,\frac{d\boldsymbol{k}_1}{k_{10}}\cdot\frac{d\boldsymbol{k}_2}{k_{20}}, \tag{19.48}$$

where $\mathfrak{M}_{\mu\nu}$ is given by expression (19.7). It is not difficult to see that

$$\varepsilon_{\mu\nu\varrho\sigma}k_{1\varrho}^{*}k_{2\sigma}^{*}\eta_\mu\eta_\nu = -\varepsilon_{\mu\nu\varrho\sigma}k_{1\varrho}k_{2\sigma}. \tag{19.49}$$

Furthermore, using the relationship

$$\varepsilon_{\mu\nu\varrho\sigma}\varepsilon_{\mu\nu\varrho'\sigma'} = 2(\delta_{\varrho\varrho'}\delta_{\sigma\sigma'} - \delta_{\varrho\sigma'}\delta_{\sigma\varrho'}),$$

as well as (19.8) we find

$$\sum_{\mu, \nu} \mathfrak{M}_{\mu\nu}\mathfrak{M}^*_{\mu\nu}\eta_\mu\eta_\nu = \tfrac{1}{2}m_\pi^4 \,|\, a \,|^2. \tag{19.50}$$

We shall obtain the total decay probability. Due to the indistinguishability of the photons the quantities $d\Gamma(k_1, k_2)$ and $d\Gamma(k_2, k_1) = d\Gamma(k_1, k_2)$ are the probability of the same event. This means that to obtain the total decay probability it is necessary to divide the result of integration over all k_1 and k_2 by 2. Setting $m_l = 0$ in (17.25) we have

$$\int \delta(k_1+k_2-q)\frac{dk_1}{k_{10}}\cdot\frac{dk_2}{k_{20}} = 2\pi. \tag{19.51}$$

Ultimately from (19.48), (19.50), and (19.51) we find the following expression for the total decay probability of the neutral π-meson into two photons:

$$\Gamma = \tfrac{1}{4}\pi\alpha^2 m_\pi^3 \,|\, a \,|^2, \tag{19.52}$$

where

$$\alpha = e^2/4\pi \simeq 1/137.$$

APPENDIX

SOLUTION OF THE FREE DIRAC EQUATION

The Covariant Density Matrix for Particles with Spin $\frac{1}{2}$

A particle's spin state is described in the general case by a density matrix. In this Appendix we shall examine in detail the covariant density matrix for particles with spin $\frac{1}{2}$. The two-component theory of the neutrino will also be examined.

We begin with the solution of the free Dirac equation

$$\alpha_k \frac{1}{i} \frac{\partial \psi}{\partial x_k} + m\beta\psi = i \frac{\partial \psi}{\partial x_0}. \tag{1}$$

Here α_k ($k = 1, 2, 3$) and β are Hermitian 4×4 matrices satisfying the commutation relations

$$\left.\begin{aligned} \alpha_i\alpha_k+\alpha_k\alpha_i &= 2\delta_{ik}, \\ \alpha_k\beta+\beta\alpha_k &= 0, \end{aligned}\right\} \tag{2}$$
$$\beta^2 = 1$$

and $\psi(x)$ is a four-component spinor. We are interested in the solutions of equation (1) which describe states with definite momentum. Such solutions have the form

$$\psi(x) = \frac{1}{(2\pi)^{3/2}} u(\boldsymbol{p})\, e^{i\boldsymbol{p}\boldsymbol{x}-iEx_0}. \tag{3}$$

Inserting (3) into (1) we find that the spinor $u(\boldsymbol{p})$ satisfies the equation

$$(\boldsymbol{\alpha}\boldsymbol{p}+m\beta)\, u(\boldsymbol{p}) = Eu(\boldsymbol{p}). \tag{4}$$

From the commutation relations (2) it follows that this equation has solutions for

$$E = \pm p_0, \tag{5}$$

where

$$p_0 = +\sqrt{m^2+\boldsymbol{p}^2}. \tag{6}$$

Let us find these solutions. We shall use a representation of the matrices α_k and β in which the matrix β is diagonal (the Dirac–Pauli representation). In this representation the matrices

α_k and β are

$$\alpha_k = \begin{pmatrix} 0 & \sigma_k \\ \sigma_k & 0 \end{pmatrix}, \quad \beta = \begin{pmatrix} I & 0 \\ 0 & -I \end{pmatrix}. \tag{7}$$

Here σ_k are Pauli matrices and I is the unit 2×2 matrix. We write the spinor u in the form

$$u = N \begin{pmatrix} \phi \\ \chi \end{pmatrix}, \tag{8}$$

where ϕ and χ are two-component spinors; N is the normalization constant. From (4), (7), and (8) we obtain the system of equations for ϕ and χ:

$$\boldsymbol{\sigma p}\chi = (E-m)\,\phi, \tag{9a}$$

$$\boldsymbol{\sigma p}\phi = (E+m)\,\chi. \tag{9b}$$

We shall first find the solutions corresponding to positive energy ($E = p_0$). From (9b) we obtain

$$\chi = \frac{\boldsymbol{\sigma p}}{p_0+m}\,\phi. \tag{10}$$

Inserting (10) into (9a) we find

$$\frac{(\boldsymbol{\sigma p})^2\,\phi}{p_0+m} = (p_0-m)\,\phi.$$

This relationship is satisfied for any $\phi((\boldsymbol{\sigma p}) = \boldsymbol{p}^2 = p_0 - m^2)$.

Thus, the solution of equation (4) corresponding to positive energy ($E = p_0$) has the form

$$u_+(\boldsymbol{p}) = N_+(\boldsymbol{p}) \begin{pmatrix} \phi \\ \dfrac{\boldsymbol{\sigma p}}{p_0+m}\,\phi \end{pmatrix}. \tag{11}$$

Let us normalize the spinor $u_+(\boldsymbol{p})$ by the following condition:

$$(u_+(\boldsymbol{p}))^+\,u_+(\boldsymbol{p}) = 1. \tag{12}$$

From (11) we obtain

$$N_+^2(\boldsymbol{p})\,(\phi^+\phi)\,\frac{2p_0}{(p_0+m)} = 1.$$

Assume

$$(\phi^+\phi) = 1. \tag{13}$$

Thus, the normalized solution of the Dirac equation corresponding to positive energy in the Dirac–Pauli representation has the form

$$u_+(\boldsymbol{p}) = \left(\frac{p_0+m}{2p_0}\right)^{1/2} \begin{pmatrix} \phi \\ \dfrac{\boldsymbol{\sigma p}}{p_0+m}\,\phi \end{pmatrix}, \tag{14}$$

where ϕ is the spinor satisfying the normalization condition (13).

We now find the solutions corresponding to negative energy ($E = -p_0$). From (9a) we obtain

$$\phi = \frac{-\boldsymbol{\sigma p}}{p_0+m}\chi. \tag{15}$$

If (15) is inserted into (9b), it is not difficult to verify that no restrictions on the spinor χ arise. Requiring that the spinor $u_-(\boldsymbol{p})$, which describes the state with negative energy, satisfy the normalization condition

$$u_-^+(\boldsymbol{p})\, u_-(\boldsymbol{p}) = 1, \tag{16}$$

and assuming that

$$\chi^+\chi = 1, \tag{17}$$

we obtain

$$u_-(\boldsymbol{p}) = \left(\frac{p_0+m}{2p_0}\right)^{1/2} \begin{pmatrix} \dfrac{-\boldsymbol{\sigma p}}{p_0+m}\chi \\ \chi \end{pmatrix}. \tag{18}$$

Obviously the spinors $u_+(\boldsymbol{p})$ and $u_-(\boldsymbol{p})$ are orthogonal, i.e.

$$u_+^+(\boldsymbol{p})\; u_-(\boldsymbol{p}) = 0. \tag{19}$$

Let us examine the Hermitian operator Σ which in the Dirac–Pauli representation is equal to

$$\Sigma_i = \begin{pmatrix} \sigma_i & 0 \\ 0 & \sigma_i \end{pmatrix}. \tag{20}$$

The operator $\frac{1}{2}\Sigma_i$ satisfies the angular momentum commutation relations

$$[\tfrac{1}{2}\Sigma_i,\ \tfrac{1}{2}\Sigma_k] = i\varepsilon_{ikl}\tfrac{1}{2}\Sigma_l \tag{21}$$

and is a spin operator.

Let us calculate the commutator of the operator Σ_i with the operator

$$H(\boldsymbol{p}) = \boldsymbol{\alpha p} + m\beta. \tag{22}$$

From (7) and (20) it is obvious that the operator Σ_i can be written in the form

$$\Sigma_i = \frac{-i}{2}\varepsilon_{ikl}\alpha_k\alpha_l. \tag{23}$$

Using commutation relations (2) we obtain

$$\left.\begin{aligned} \alpha_k\alpha_l\beta - \beta\alpha_k\alpha_l &= \alpha_k(\alpha_l\beta+\beta\alpha_l) = 0, \\ \alpha_k\alpha_l\alpha_m - \alpha_m\alpha_k\alpha_l &= 2\alpha_k\,\delta_{lm} - 2\alpha_l\,\delta_{km}. \end{aligned}\right\} \tag{24}$$

Thus,

$$[\Sigma_i,\, H(\boldsymbol{p})] = -2i(\boldsymbol{\alpha}\times\boldsymbol{p})_i. \tag{25}$$

Hence, the operator of the spin projection on the momentum commutes with the operator $H(\boldsymbol{p})$ (the remaining components of the spin operator are not conserved).

Denote by $\boldsymbol{n}$ the unit vector in the direction of the momentum ($\boldsymbol{n} = \boldsymbol{p}/|\boldsymbol{p}|$). The operator

$$\Sigma \boldsymbol{n} \tag{26}$$

is called the *helicity operator*. We shall require that the spinors $u_+(\boldsymbol{p})$ and $u_-(\boldsymbol{p})$, which are solutions of Dirac's equation (4), also be eigenfunctions of the helicity operator. We have

$$\Sigma \boldsymbol{n} u_+^r(\boldsymbol{p}) = r u_+^r(\boldsymbol{p}), \quad \Sigma \boldsymbol{n} u_-^r(\boldsymbol{p}) = r u_-^r(\boldsymbol{p}). \tag{27}$$

Obviously $r = \pm 1$. The spinors $u_+^r(\boldsymbol{p})$ and $u_-^r(\boldsymbol{p})$ describe the spin state of Dirac particles with energy p_0 and $-p_0$ and helicity r. From (12), (16), (19), and (27) we have

$$\left.\begin{aligned} (u_+^r(\boldsymbol{p}))^+ \, u_+^{r'}(\boldsymbol{p}) &= \delta_{rr'}, \\ (u_-^r(\boldsymbol{p}))^+ \, u_-^{r'}(\boldsymbol{p}) &= \delta_{rr'}, \\ (u_+^r(\boldsymbol{p}))^+ \, u_-^{r'}(\boldsymbol{p}) &= 0. \end{aligned}\right\} \tag{28}$$

The functions

$$\frac{1}{(2\pi)^{3/2}} u_+^r(\boldsymbol{p}) \, e^{i\boldsymbol{px} - ip_0x_0}, \quad \frac{1}{(2\pi)^{3/2}} u_-^r(\boldsymbol{p}) \, e^{i\boldsymbol{px} + ip_0x_0} \qquad (r = \pm 1)$$

form a complete set of solutions of the Dirac equation for definite momentum $\boldsymbol{p}$. We write the spinors $u_+^r(\boldsymbol{p})$ in the form

$$\left.\begin{aligned} u_+^r(\boldsymbol{p}) &= \left(\frac{p_0+m}{2p_0}\right)^{1/2} \begin{pmatrix} \phi^r \\ \dfrac{\boldsymbol{\sigma p}}{p_0+m} \phi^r \end{pmatrix}, \\ u_-^r(\boldsymbol{p}) &= \left(\frac{p_0+m}{2p_0}\right)^{1/2} \begin{pmatrix} \dfrac{-\boldsymbol{\sigma p}}{p_0+m} \chi^r \\ \chi^r \end{pmatrix}. \end{aligned}\right\} \tag{29}$$

From (27) we obtain

$$\left.\begin{aligned} \boldsymbol{\sigma n} \phi^r &= r\phi^r, \\ \boldsymbol{\sigma n} \chi^r &= r\chi^r. \end{aligned}\right\} \tag{30}$$

Thus, equations (27) determine the choice of a two-component spinor ϕ for states with positive energy and a spinor χ for states with negative energy.

From (29) and (30) we have

$$u_+^r(\boldsymbol{p}) = \left(\frac{p_0+m}{2p_0}\right)^{1/2} \begin{pmatrix} \phi^r \\ \dfrac{|\boldsymbol{p}|\, r}{p_0+m} \phi^r \end{pmatrix},$$

$$u_-^r(\boldsymbol{p}) = \left(\frac{p_0+m}{2p_0}\right)^{1/2} \begin{pmatrix} \dfrac{-|\boldsymbol{p}|\, r\chi^r}{p_0+m} \\ \chi^r \end{pmatrix}:$$

From this we find for $m = 0$ (the neutrino)

$$u_+^r(\boldsymbol{p}) = \frac{1}{\sqrt{2}} \begin{pmatrix} \phi^r \\ r\phi^r \end{pmatrix}, \quad u_-^r(\boldsymbol{p}) = \frac{1}{\sqrt{2}} \begin{pmatrix} -r\chi^r \\ \chi^r \end{pmatrix}. \tag{31}$$

It is not difficult to see that the spinors (31) are eigenstates of the matrix γ_5. Actually, in the Dirac–Pauli representation the matrix γ_5 has the form

$$\gamma_5 = i\alpha_1\alpha_2\alpha_3 = \begin{pmatrix} 0 & - \\ -I & 0 \end{pmatrix}. \tag{32}$$

From (31) and (32) we obtain

$$\left.\begin{aligned} \gamma_5 u_+^{(-1)}(\boldsymbol{p}) &= u_+^{(-1)}(\boldsymbol{p}), \\ \gamma_5 u_-^{(1)}(\boldsymbol{p}) &= u_-^{(1)}(\boldsymbol{p}), \\ \gamma_5 u_+^{(1)}(\boldsymbol{p}) &= -u_+^{(1)}(\boldsymbol{p}), \\ \gamma_5 u_-^{(-1)}(\boldsymbol{p}) &= -u_-^{(-1)}(\boldsymbol{p}). \end{aligned}\right\} \tag{33}$$

Thus, in the case of the neutrino the states $u_+^{(-1)}(\boldsymbol{p})$ and $u_-^{(1)}(\boldsymbol{p})$ ($u_+^{(1)}(\boldsymbol{p})$ and $u_-^{(-1)}(\boldsymbol{p})$) are eigenstates of the matrix γ_5 corresponding to the eigenvalue $+1$ (-1).

The spinors $u_+(\boldsymbol{p})$ and $u_-(\boldsymbol{p})$ are normalized by conditions (12) and (16). The quantity

$$u^+(\boldsymbol{p})\, u(\boldsymbol{p}) = \bar{u}(\boldsymbol{p})\, \gamma_4 u(\boldsymbol{p})$$

is the fourth component of the vector $\bar{u}(\boldsymbol{p})\, \gamma_\mu u(\boldsymbol{p})$ and consequently this normalization is not invariant. We shall introduce invariantly normalized spinors. Let us examine particles with non-zero mass. We obtain from equation (4) by Hermitian conjugation

$$u^+(\boldsymbol{p})(\boldsymbol{\alpha p}+m\beta) = u^+(\boldsymbol{p})\, E. \tag{34}$$

We multiply (4) on the left by $u^+(\boldsymbol{p})\,\beta$ and (34) on the right by $\beta u(\boldsymbol{p})$. Adding the resulting equations and using the commutation relations (2) we find

$$\bar{u}(\boldsymbol{p})\, u(\boldsymbol{p}) = \frac{m}{E}\, u^+(\boldsymbol{p})\, u(\boldsymbol{p}). \tag{35}$$

Since $u^+(\boldsymbol{p})\, u(\boldsymbol{p}) \geqslant 0$, it is obvious from relations (35) that

$$\bar{u}_+(\boldsymbol{p})\, u_+(\boldsymbol{p}) \geqslant 0 \quad \text{and} \quad \bar{u}_-(\boldsymbol{p})\, u_-(\boldsymbol{p}) \leqslant 0.$$

Note also that for particles with zero mass, $\bar{u}(\boldsymbol{p})\, u(\boldsymbol{p}) = 0$.

Let us first examine states with positive energy. We introduce the spinor $u(p)$ related to $u_+(\boldsymbol{p})$ by the relationship

$$u(p) = \left(\frac{p_0}{m}\right)^{1/2} u_+(\boldsymbol{p}). \tag{36}$$

From (35) and (28) we obtain

$$\bar{u}(p)\, u(p) = \frac{p_0}{m}\, \bar{u}_+(\boldsymbol{p})\, u_+(\boldsymbol{p}) = 1. \tag{37}$$

Thus, the spinor

$$u(p) = \left(\frac{p_0+m}{2m}\right)^{1/2} \begin{pmatrix} \phi \\ \dfrac{\boldsymbol{\sigma p}}{p_0+m}\,\phi \end{pmatrix} \tag{38}$$

is normalized by the invariant condition (37).

Now we examine the states with negative energy. Define the spinor $u(-p)$:

$$u(-p) = \left(\frac{p_0}{m}\right)^{1/2} u_-(-\boldsymbol{p}). \tag{39}$$

From (35) and (28) we obtain

$$\bar{u}(-p)\,u(-p) = \frac{p_0}{m}\,\bar{u}_-(-\boldsymbol{p})\,u_-(-\boldsymbol{p}) = -1. \tag{40}$$

The spinor

$$u(-p) = \left(\frac{p_0+m}{2m}\right)^{1/2} \begin{pmatrix} \dfrac{\boldsymbol{\sigma p}}{p_0+m}\,\chi \\ \chi \end{pmatrix}, \tag{41}$$

which describes the spin state of a particle with energy $-p_0$ and momentum $-\boldsymbol{p}$, is also normalized invariantly.

The spinors $u(p)$ and $u(-p)$ satisfy the equations

$$(\boldsymbol{\alpha p}+m\beta)\,u(p) = p_0 u(p), \tag{42a}$$

$$(-\,\boldsymbol{\alpha p}+m\beta)\,u(-p) = -p_0 u(-p). \tag{42b}$$

We multiply these equations on the left by the matrix $-i\beta$. We obtain

$$(\hat{p}-im)\,u(p) = 0, \tag{43a}$$

$$(\hat{p}+im)\,u(-p) = 0, \tag{43b}$$

where

$$\left.\begin{aligned} \gamma_k &= -i\beta\alpha_k = i\alpha_k\beta, \qquad k = 1, 2, 3, \\ \gamma_4 &= \beta, \\ \hat{p} &= p_\mu \gamma_\mu. \end{aligned}\right\} \tag{44}$$

From the commutation relations (2) it is obvious that the matrices γ_μ satisfy the relationships

$$\gamma_\mu\gamma_\nu+\gamma_\nu\gamma_\mu = 2\delta_{\mu\nu}. \tag{45}$$

It is also clear that $\gamma_\mu^+ = \gamma_\mu$. We find by Hermitian conjugation of (43)

$$u^+(p)(p_\mu^*\gamma_\mu+im) = 0.$$

Multiplying this equation on the right by γ_4 and recognizing that

$$\gamma_4 p_\mu^* \gamma_\mu \gamma_4 = -\gamma_\mu p_\mu,$$

we obtain

$$\bar{u}(p)(\hat{p}-im) = 0. \tag{46a}$$

Analogously we find

$$\bar{u}(-p)(\hat{p}+im) = 0. \tag{46b}$$

Multiplying equation (46b) on the right by $u(p)$ and equation (43a) on the left by $\bar{u}(-p)$ and adding the resulting relationships, we find

$$\bar{u}(-p)\,u(p) = 0, \tag{47}$$

i.e.

$$\bar{u}_-(-\boldsymbol{p})\,u_+(\boldsymbol{p}) = 0.$$

Note that

$$\bar{u}_-(\boldsymbol{p})\,u_+(\boldsymbol{p}) \neq 0.$$

If the spinors $u^r_+(\boldsymbol{p})$ and $u^r_-(-\boldsymbol{p})$ are eigenfunctions of the helicity operator (26), i.e.

$$\left.\begin{aligned} \boldsymbol{\Sigma}\,\boldsymbol{n}u^r_+(\boldsymbol{p}) &= ru^r_+(\boldsymbol{p}),\\ \boldsymbol{\Sigma}(-\boldsymbol{n})\,u^r_-(-\boldsymbol{p}) &= ru^r_-(-\boldsymbol{p}) \end{aligned}\right\} \tag{48}$$

($\boldsymbol{n} = \boldsymbol{p}/|\boldsymbol{p}|$), then it is obvious that the spinors

$$u^r(p) = \left(\frac{p_0}{m}\right)^{1/2} u^r_+(\boldsymbol{p}) \quad \text{and} \quad u^r(-p) = \left(\frac{p_0}{m}\right)^{1/2} u^r_-(-\boldsymbol{p})$$

also describe states with definite helicity. Such spinors satisfy the following conditions of normalization and orthogonality:

$$\left.\begin{aligned} \bar{u}^{r'}(p)\,u^r(p) &= \delta_{rr'},\\ \bar{u}^{r'}(-p)\,u^r(-p) &= -\delta_{r'r},\\ \bar{u}^{r'}(-p)\,u^r(p) &= 0. \end{aligned}\right\} \tag{49}$$

From (38) in the rest system ($m \neq 0$) for the state with positive energy we obtain

$$u(im) = \begin{pmatrix}\phi\\0\end{pmatrix}. \tag{50}$$

Let us perform a Lorentz transformation from the rest system to the system where the particle's momentum is $\boldsymbol{p}$. We direct the 1 axis along the momentum $\boldsymbol{p}$. We have

$$\left.\begin{aligned} u(p) &= L(\boldsymbol{p})\,u(im),\\ \bar{u}(p) &= \bar{u}(im)\,L^{-1}(\boldsymbol{p}). \end{aligned}\right\} \tag{51}$$

The matrix $L(\boldsymbol{p})$ satisfies the relationships

$$L^{-1}(\boldsymbol{p})\,\gamma_\mu L(\boldsymbol{p}) = a_{\mu\nu}\gamma_\nu, \tag{52}$$

where

$$\left.\begin{aligned} &a_{11} = \frac{p_0}{m}, \qquad a_{14} = -i\frac{|\boldsymbol{p}|}{m},\\ &a_{41} = i\frac{|\boldsymbol{p}|}{m}, \qquad a_{44} = \frac{p_0}{m}, \qquad a_{22} = 1, \qquad a_{33} = 1. \end{aligned}\right\} \tag{53}$$

The remaining components $a_{\mu\nu}$ are zero. Note that

$$a_{\mu\nu}a_{\mu\nu'} = \delta_{\nu\nu'}, \quad a_{\mu\nu}a_{\mu'\nu} = \delta_{\mu\mu'}.$$

Let us find the matrix $L(\boldsymbol{p})$. For this we introduce the quantity α which is defined by

$$\tanh \alpha = \frac{|\boldsymbol{p}|}{p_0}. \tag{54}$$

Obviously

$$\frac{p_0}{m} = \cosh \alpha, \qquad \frac{|\boldsymbol{p}|}{m} = \sinh \alpha.$$

Using (52)–(54) we obtain

$$L^{-1}(\boldsymbol{p})\,\gamma_\mu L(\boldsymbol{p}) = \left(\cosh \frac{\alpha}{2} + i\gamma_1\gamma_4 \sinh \frac{\alpha}{2}\right) \gamma_\mu \left(\cosh \frac{\alpha}{2} - i\gamma_1\gamma_4 \sinh \frac{\alpha}{2}\right), \tag{55}$$

from which

$$\left.\begin{aligned} L(\boldsymbol{p}) &= \cosh \frac{\alpha}{2} - i\gamma_1\gamma_4 \sinh \frac{\alpha}{2}, \\ L^{-1}(\boldsymbol{p}) &= \cosh \frac{\alpha}{2} + i\gamma_1\gamma_4 \sinh \frac{\alpha}{2}. \end{aligned}\right\} \tag{56}$$

It is not difficult to see that

$$\cosh \frac{\alpha}{2} = \sqrt{\frac{p_0+m}{2m}}, \qquad \sinh \frac{\alpha}{2} = \frac{|\boldsymbol{p}|}{\sqrt{2m(p_0+m)}}. \tag{57}$$

From (56) and (57) we find that the matrices $L(\boldsymbol{p})$ and $L^{-1}(\boldsymbol{p})$ are equal to

$$\left.\begin{aligned} L(\boldsymbol{p}) &= \sqrt{\frac{p_0+m}{2m}} \left(1 - i\gamma_1\gamma_4 \frac{|\boldsymbol{p}|}{p_0+m}\right), \\ L^{-1}(\boldsymbol{p}) &= \sqrt{\frac{p_0+m}{2m}} \left(1 + i\gamma_1\gamma_4 \frac{|\boldsymbol{p}|}{p_0+m}\right). \end{aligned}\right\} \tag{58}$$

Since the 1 axis is directed along $\boldsymbol{p}$, then

$$\gamma_1 |\boldsymbol{p}| = \boldsymbol{\gamma}\boldsymbol{p}.$$

We write the final matrix $L(\boldsymbol{p})$ in the form:

$$L(\boldsymbol{p}) = \left(\frac{p_0+m}{2m}\right)^{1/2} \left(1 - i\frac{\boldsymbol{\gamma}\boldsymbol{p}}{p_0+m}\gamma_4\right). \tag{59}$$

In the Dirac–Pauli representation we obtain from (50), (51), and (59)

$$u(p) = L(\boldsymbol{p})\, u(im) = \left(\frac{p_0+m}{2m}\right)^{1/2} \begin{pmatrix} \phi \\ \dfrac{\boldsymbol{\sigma}\boldsymbol{p}}{p_0+m}\phi \end{pmatrix}, \tag{60}$$

which coincides with (38). From this examination the physical meaning of the spinor ϕ is obvious. The spinor ϕ describes the spin state of a particle in the rest system whose axes coincide with the axes of the system where the particle's momentum is $\boldsymbol{p}$.

The state with definite helicity is given by equation (30). The spinor $\phi^{(r)}$ which satisfies this equation describes the state in the rest system with the same spin projection r in the direction of the momentum (in the rest system the direction is opposite the velocity of the system in which the particle's momentum is $\boldsymbol{p}$). It is clear that the spin state of the particle can be obtained if we require that the spinor ϕ be an eigenfunction of the spin projection operator in any fixed direction in the rest system.

Let the spinor ϕ^s satisfy the equation

$$\boldsymbol{\sigma n}^0\phi^s = s\phi^s, \tag{61}$$

where $\boldsymbol{n}^0$ is a unit vector in the rest system. Obviously, $s = \pm 1$. From (61) by Hermitian conjugation we obtain

$$\phi^{s+}\boldsymbol{\sigma n}^0 = \phi^{s+}s. \tag{62}$$

Multiplying (61) on the left by $\phi^{s+}\sigma$ and (62) on the right by $\sigma_i\phi^s$, adding the resulting equations and using the commutation relations for the matrices σ_i, we obtain

$$\phi^{s+}\boldsymbol{\sigma}\phi^s = s\boldsymbol{n}^0. \tag{63}$$

Thus, $s\boldsymbol{n}^0$ is the mean value of the operator σ in the state ϕ^s. Equation (61) can be written

$$\boldsymbol{\Sigma n}^0 u^s(im) = su^s(im). \tag{64}$$

It is not difficult to see that

$$\boldsymbol{\Sigma} = i\gamma_5\boldsymbol{\gamma}\gamma_4. \tag{65}$$

From (64) and (65), using $\gamma_4 u^s(im) = u^s(im)$, we obtain

$$i\gamma_5\boldsymbol{\gamma n}^0 u^s(im) = su^s(im). \tag{66}$$

This equation can be written

$$i\gamma_5\gamma n^0 u^s(im) = su^s(im), \tag{67}$$

where

$$n^0 = (\boldsymbol{n}^0, 0). \tag{68}$$

Multiply (67) by the matrix $L(\boldsymbol{p})$. Using (51) we obtain

$$L(\boldsymbol{p})\, i\gamma_5\gamma_\nu L^{-1}(\boldsymbol{p})\, n_\nu^0 u^s(p) = su^s(p). \tag{69}$$

With the help of (52) and (53) we find (for the transformation we are examining $\det a = 1$)

$$L(\boldsymbol{p})\,\gamma_5\gamma_\nu L^{-1}(\boldsymbol{p}) = a_{\mu\nu}\gamma_5\gamma_\mu. \tag{70}$$

From (69) and (70) we obtain

$$i\gamma_5\gamma n u^s(p) = su^s(p), \tag{71}$$

where

$$n_\mu = a_{\mu\nu}n_\nu^0. \tag{72}$$

The quantities n_μ, which in the rest system were n_μ^0, form a four-vector. Obviously n_μ is a space-like four-vector (in the rest system its fourth component is zero). Since $(\boldsymbol{n}^0)^2 = 1$, it is clear that $n^2 = 1$. Furthermore,

$$np = n_4^0 im = 0. \tag{73}$$

Thus, the spinor $u^s(p)$, which is an eigenfunction of the operator $\hat{p}$ and corresponds to the eigenvalue im, is also an eigenfunction of the operator $i\gamma_5\hat{n}$ corresponding to the eigenvalue s ($s = \pm 1$). It is not difficult to verify that the operators $\hat{p}$ and $\gamma_5\hat{n}$ commute. Actually,

$$\hat{p}\gamma_5\hat{n} - \gamma_5\hat{n}\hat{p} = -\gamma_5(\hat{p}\hat{n} + \hat{n}\hat{p}) = -2\gamma_5 pn = 0.$$

From (71) by Hermitian conjugation and multiplication on the right by γ_4 we obtain

$$\bar{u}^s(p)\, i\gamma_5\hat{n} = s\bar{u}^s(p). \tag{74}$$

We multiply (74) on the right by $i\gamma_5\gamma_\mu u^s(p)$ and (71) on the left by $\bar{u}^s(p) i\gamma_5\gamma_\mu$. Adding the resulting relationships we obtain

$$\bar{u}^s(p)(\hat{n}\gamma_\mu + \gamma_\mu\hat{n})\, u^s(p) = 2s\bar{u}^s(p)\, i\gamma_5\gamma_\mu u^s(p).$$

From this, using the commutation relations for the γ-matrices and the normalization condition (37), we find

$$\bar{u}^s(p)\, i\gamma_5\gamma_\mu u^s(p) = sn_\mu. \tag{75}$$

Thus, sn_μ is the mean value of the operator $i\gamma_5\gamma_\mu$ in the state described by the spinor $u^s(p)$.

Up to this point we have assumed that the spin state is described by a wave function. In the general case, however, the spin state of the particles is described by a density matrix. The covariant density matrix for a relativistic particle with spin $\frac{1}{2}$ is given by the expression

$$\varrho_{\sigma\sigma'}(p) = \sum_s u_\sigma^s(p)\, \bar{u}_{\sigma'}^s(p)\, \alpha_s. \tag{76}$$

Here $u^s(p)$ are the normalized eigenfunctions of the operator $i\gamma_5\hat{n}$ ($np = 0$, $n^2 = 1$); α_s are real positive quantities [α_s is the weight with which the function $u^s(p)$ enters the mixed state described by the density matrix (76)]. The mean value of the operator O is

$$\langle O \rangle = \frac{\sum\limits_{s, \sigma', \sigma} \bar{u}_{\sigma'}^s(p)\, O_{\sigma'\sigma} u_\sigma^s(p)\, \alpha_s}{\Sigma\alpha_s} = \frac{\mathrm{Tr}\,[O\varrho(p)]}{\mathrm{Tr}\,[\varrho(p)]}. \tag{77}$$

Let us ascertain the conditions satisfied by the density matrix $\varrho(p)$. If the state is described by a wave function (a pure state), the density matrix is

$$\varrho_{\sigma\sigma'}^0(p) = u_\sigma(p)\, \bar{u}_{\sigma'}(p). \tag{78}$$

Such a density matrix satisfies the condition

$$\mathrm{Tr}\,[\varrho^0(p)]^2 = \sum_{\sigma, \sigma'} u_\sigma(p)\, \bar{u}_{\sigma'}(p)\, u_{\sigma'}(p)\, \bar{u}_\sigma(p) = \{\mathrm{Tr}\,[\varrho^0(p)]\}^2. \tag{79}$$

In the general case we obtain from (76)

$$\mathrm{Tr}\,[\varrho(p)]^2 = \sum_{s, s', \sigma, \sigma'} u_\sigma^s(p)\, \bar{u}_{\sigma'}^s(p)\, u_{\sigma'}^{s'}(p)\, \bar{u}_\sigma^{s'}(p)\, \alpha_s\alpha_{s'} = \sum_s \alpha_s^2. \tag{80}$$

Since $\alpha_s \geqslant 0$, it is obvious that

$$\sum_s \alpha_s^2 \leqslant \Big(\sum_s \alpha_s\Big)^2. \tag{81}$$

Considering that $\mathrm{Tr}\, \varrho = \sum_s \alpha_s$, from (80) and (81) we conclude that in the general case the density matrix $\varrho(p)$ satisfies the condition

$$\mathrm{Tr}\, [\varrho(p)]^2 \leqslant \{\mathrm{Tr}\, [\varrho(p)]\}^2. \tag{82}$$

One can show that in the case when $\mathrm{Tr}\, \varrho^2 = (\mathrm{Tr}\, \varrho)^2$, the density matrix has the form (78), i.e. the spin state is described by one wave function.

Furthermore, since the coefficients α_s are real we have

$$\bar{\varrho}(p) = \varrho(p), \tag{83}$$

where

$$\bar{\varrho}(p) = \gamma_4 \varrho^+(p)\, \gamma_4.$$

We introduce the operator

$$\Lambda(p) = \frac{\hat{p}+im}{2im}, \tag{84}$$

$$\Lambda(-p) = \frac{-\hat{p}+im}{2im}. \tag{85}$$

Using equations (43) and (46) we obtain

$$\left.\begin{aligned} \Lambda(p)\, u(p) &= u(p), \\ \bar{u}(p)\, \Lambda(p) &= \bar{u}(p), \\ \Lambda(p)\, u(-p) &= 0, \\ \bar{u}(-p)\, \Lambda(p) &= 0 \end{aligned}\right\} \tag{86}$$

and

$$\left.\begin{aligned} \Lambda(-p)\, u(-p) &= u(-p), \\ \bar{u}(-p)\, \Lambda(-p) &= \bar{u}(-p), \\ \Lambda(-p)\, u(p) &= 0, \\ \bar{u}(p)\, \Lambda(-p) &= 0. \end{aligned}\right\} \tag{87}$$

From (86) and (87) it follows that $\Lambda(p)$ and $\Lambda(-p)$ are projection operators. It is easy to see that

$$\left.\begin{aligned} \Lambda(p)+\Lambda(-p) &= 1, \\ \Lambda(p)\, \Lambda(-p) = \Lambda(-p)\, \Lambda(p) &= 0, \\ \Lambda(p)\, \Lambda(p) &= \Lambda(p), \\ \Lambda(-p)\, \Lambda(-p) &= \Lambda(-p). \end{aligned}\right\} \tag{88}$$

The density matrix (76) describes the spin state of a particle with positive energy p_0 and momentum $\boldsymbol{p}$. From (76) and (87) it is obvious that the matrix $\varrho(p)$ satisfies the conditions

$$\Lambda(-p)\, \varrho(p) = 0, \quad \varrho(p)\, \Lambda(-p) = 0. \tag{89}$$

With the help of the first of the relationships (88) these conditions can also be written in the forms

$$\Lambda(p)\,\varrho(p) = \varrho(p), \quad \varrho(p)\,\Lambda(p) = \varrho(p). \tag{90}$$

Note finally that the normalization condition of the density matrix has the form

$$\text{Tr}\,[\varrho(p)] = 1. \tag{91}$$

We now construct the general expression for the normalized density matrix of a relativistic particle with spin $\frac{1}{2}$. The matrix $\varrho(p)$ always can be expanded into sixteen Dirac matrices. We obtain

$$\varrho(p) = a + b_\nu\gamma_\nu + c_{\varrho\sigma}\sigma_{\varrho\sigma} + id_\nu\gamma_5\gamma_\nu + e\gamma_5$$
$$\left(\sigma_{\varrho\sigma} = \frac{1}{2i}(\gamma_\varrho\gamma_\sigma - \gamma_\sigma\gamma_\varrho)\right). \tag{92}$$

From the normalization condition (91) we find

$$\text{Tr}\,\varrho = 1 = 4a, \quad a = \tfrac{1}{4}. \tag{93}$$

Multiplying (92) by γ_μ and calculating the traces of both sides of the resulting equation, we obtain

$$4b_\mu = \text{Tr}\,[\gamma_\mu\varrho]. \tag{94}$$

Furthermore, with the help of (90) we find

$$b_\mu = \tfrac{1}{4}\,\text{Tr}\,[\gamma_\mu\varrho\Lambda(p)] = \tfrac{1}{4}\,\text{Tr}\,[\gamma_\mu\Lambda(p)\,\varrho\Lambda(p)]. \tag{95}$$

Using the commutation relations for the matrices γ_μ, it is easy to see that

$$\gamma_\mu\Lambda(p) = \Lambda(-p)\,\gamma_\mu + \frac{p_\mu}{im}. \tag{96}$$

Inserting this relationship into (95) and using (88) and (91) we find

$$b_\mu = \frac{p_\mu}{4im}. \tag{97}$$

We multiply (92) by γ_5. By calculating the trace we obtain

$$4e = \text{Tr}\,[\varrho\gamma_5] = \text{Tr}\,[\Lambda(p)\,\varrho\Lambda(p)\,\gamma_5] = \text{Tr}\,[\Lambda(p)\,\varrho\gamma_5\Lambda(-p)] - 0. \tag{98}$$

Define the quantity

$$\xi_\mu = \text{Tr}\,(i\gamma_5\gamma_\mu\varrho). \tag{99}$$

It is not difficult to see that ξ_μ transforms like a pseudovector. Furthermore, since

$$\hat{p}\Lambda(p) = im\Lambda(p),$$

we obtain

$$\xi p = \text{Tr}\,[i\gamma_5\,\hat{p}\Lambda(p)\,\varrho] = -m\,\text{Tr}\,(\gamma_5\varrho) = 0. \tag{100}$$

From (92) we find that

$$d_\mu = \tfrac{1}{4}\xi_\mu. \tag{101}$$

It is not difficult to verify the validity of the relationship

$$\sigma_{\varrho\sigma} = -\tfrac{1}{2}\varepsilon_{\varrho\sigma\mu\nu}\sigma_{\mu\nu}\gamma_5. \tag{102}$$

With the help of (102) we write the third term of the expansion (92) in the form

$$c_{\varrho\sigma}\sigma_{\varrho\sigma} = c'_{\mu\nu}\sigma_{\mu\nu}\gamma_5. \tag{103}$$

Obviously

$$c'_{\mu\nu} = -c'_{\nu\mu}. \tag{104}$$

To find $c'_{\mu\nu}$ we multiply (92) by $\sigma_{\mu\nu}\gamma_5$. Calculating the trace of both sides of the resulting equation and using

$$\operatorname{Tr}(\sigma_{\mu\nu}\gamma_5\sigma_{\mu'\nu'}\gamma_5) = 4(\delta_{\mu\mu'}\delta_{\nu\nu'} - \delta_{\mu\nu'}\delta_{\nu\mu'}), \tag{105}$$

we obtain

$$c'_{\mu\nu} = \tfrac{1}{8}\operatorname{Tr}(\varrho\sigma_{\mu\nu}\gamma_5). \tag{106}$$

Furthermore,

$$\operatorname{Tr}(\varrho\sigma_{\mu\nu}\gamma_5) = \frac{1}{i}\operatorname{Tr}[\varrho(\gamma_\mu\gamma_\nu - \delta_{\mu\nu})\gamma_5] = \frac{1}{i}\operatorname{Tr}[\Lambda(p)\,\varrho\Lambda(p)\,\gamma_\mu\gamma_\nu\gamma_5]. \tag{107}$$

It is not difficult to see that

$$\Lambda(p)\gamma_\mu\gamma_\nu\gamma_5 = \frac{p_\mu}{im}\gamma_\nu\gamma_5 - \frac{p_\nu}{im}\gamma_\mu\gamma_5 + \gamma_\mu\gamma_\nu\gamma_5\Lambda(-p). \tag{108}$$

Inserting this relationship into (107) and using (88) and (99) we obtain

$$c'_{\mu\nu} = \frac{1}{8im}(p_\mu\xi_\nu - p_\nu\xi_\mu). \tag{109}$$

Finally, inserting (93), (97), (101), (103), and (109) into expansion (92), we find for the density matrix

$$\varrho(p) = \tfrac{1}{4} + \frac{\hat{p}}{4im} + \tfrac{1}{4}i\gamma_5\hat{\xi} + \frac{1}{4im}\hat{p}i\gamma_5\hat{\xi} = \frac{\hat{p}+im}{2im}\,\tfrac{1}{2}(1+i\gamma_5\hat{\xi}). \tag{110}$$

Note that this expression can also be written as

$$\varrho(p) = \tfrac{1}{2}(1+i\gamma_5\hat{\xi})\frac{\hat{p}+im}{2im}$$

(the operators $\hat{p}$ and $\gamma_5\hat{\xi}$ commute).

The vector ξ_μ, which satisfies the condition $\xi p = 0$, characterizes the spin state of relativistic spin $\frac{1}{2}$ particles. It is called the polarization four-vector. It follows from (83) that the space components of the polarization vector are real and the fourth component is imaginary.

We write the four-vector ξ_μ in the form

$$\xi_\mu = Pn_\mu, \tag{111}$$

where n_μ is a unit space-like vector. Obviously

$$P^2 = \xi^2. \tag{112}$$

The invariant $\sqrt{P^2}$ is called the polarization. From (82) and (110) we obtain

$$\text{Tr}\, \varrho^2 = \tfrac{1}{2}(1+\xi^2) \leqslant 1,$$

from which

$$P^2 \leqslant 1. \tag{113}$$

With the help of (75), (76), and (99) we find that the polarization four-vector is

$$\xi_\mu = \sum_s \bar{u}^s(p)\, i\gamma_5\gamma_\mu u^s(p)\, \alpha_s = n_\mu \sum_s \alpha_s s.$$

Comparing this expression to (111) we conclude that

$$P = \sum_s s\alpha_s.$$

From this relationship and the normalization condition (91) we find

$$\alpha_1 = \frac{1+P}{2}, \quad \alpha_{-1} = \frac{1-P}{2}.$$

If $P = 1$, then $\alpha_{-1} = 0, \alpha_1 = 1$, and the state is described by the spinor $u^1(p)$. For $P = -1$ the state is described by the spinor $u^{-1}(p)$. In the case of an unpolarized state ($P = 0$) $\alpha_1 = \alpha_{-1} = \frac{1}{2}$. The density matrix of such a state is

$$\varrho_0(p) = \tfrac{1}{2}\sum_s u^s(p)\, \bar{u}^s(p) = \tfrac{1}{2}\Lambda(p). \tag{114}$$

From the condition $\xi p = 0$ it follows that

$$\xi_4 = \frac{i\boldsymbol{\xi}\mathbf{p}}{p_0}. \tag{115}$$

Thus, the density matrix of a relativistic spin $\frac{1}{2}$ particle is characterized by the vector ξ.

In the rest system $\xi_4^0 = 0$ and $(\boldsymbol{\xi}^0)^2 = P^2$. The polarization vector in the system in which the momentum of the particle is $\mathbf{p}$ is related to the polarization vector in the rest system by the Lorentz transformation

$$\xi_1 = \frac{\xi_1^0 p_0}{m}, \quad \xi_2 = \xi_2^0, \quad \xi_3 = \xi_3^0 \tag{116a}$$

(the 1 axis is directed along the momentum). This transformation can also be written in the form

$$\boldsymbol{\xi} = \boldsymbol{\xi}^0 + (\boldsymbol{\xi}^0\mathbf{p})\frac{\mathbf{p}}{m(p_0+m)}. \tag{116b}$$

We now discuss the case of antiparticles with spin $\frac{1}{2}$. Let us take as an example a process with a particle and an antiparticle in the initial state. Denote by p' and p the four-momenta of the particle and the antiparticle. In accordance with the general rule discussed in detail in § 11, the matrix element of the process can be written in the form

$$M_{fi} = \bar{u}^s(-p)\,\mathfrak{M}\,u^{s'}(p'). \tag{117}$$

The square of the matrix element averaged over the spin states of the initial particles is

$$\sum |M_{fi}|^2 = \sum_{s,\,s'} |\bar{u}^s(-p)\,\mathfrak{M}\,u^{s'}(p')|^2 \alpha_{s'}\,\beta_s = \mathrm{Tr}\,[\mathfrak{M}\,\varrho(p')\,\overline{\mathfrak{M}\,\varrho(-p)}]. \tag{118}$$

Here

$$\varrho_{\sigma\sigma'}(-p) = \sum_s u^s_\sigma(-p)\,\bar{u}^s_{\sigma'}(-p)\,\beta_s \tag{119}$$

and $\varrho(p')$ is the density matrix of the particle with momentum p'. The matrix $\varrho(-p)$ characterizes the spin state of the initial antiparticle with momentum p (β_s is the probability that the initial antiparticle is found in the state described by the spinor $u^s(p)$). The normalization condition is

$$\sum_s \beta_s = 1. \tag{120}$$

Using $\bar{u}^s(-p)\,u^s(-p) = -1$, we obtain from (119) and (120) the normalization condition for the matrix $\varrho(-p)$

$$\mathrm{Tr}\,\varrho(-p) = -1. \tag{121}$$

With the help of (119) we find that the matrix $\varrho(-p)$ obeys the following conditions analogous to (90):

$$\left.\begin{aligned} \Lambda(-p)\,\varrho(-p) &= \varrho(-p), \\ \varrho(-p)\,\Lambda(-p) &= \varrho(-p). \end{aligned}\right\} \tag{122}$$

To find the general expression for the matrix $\varrho(-p)$ we expand it into sixteen basis matrices. We obtain

$$\varrho(-p) = A + B_\mu\gamma_\mu + C_{\mu\nu}\sigma_{\mu\nu}\gamma_5 + D_\mu i\gamma_5\gamma_\mu + E\gamma_5. \tag{123}$$

From (121) we find

$$A = -\tfrac{1}{4}. \tag{124}$$

Furthermore, we obtain analogous to (97) and (98),

$$\left.\begin{aligned} B_\mu &= \tfrac{1}{4}\,\mathrm{Tr}\,[\varrho(-p)\,\gamma_\mu] = -\frac{p_\mu}{4im}\,\mathrm{Tr}\,[\varrho(-p)] = \frac{p_\mu}{4im}, \\ E &= \tfrac{1}{4}\,\mathrm{Tr}\,[\varrho(-p)\,\gamma_5] = 0. \end{aligned}\right\} \tag{125}$$

The antiparticle polarization vector (the mean value of the operator $i\gamma_5\gamma_\mu$) is given by the expression

$$\xi_\mu = \frac{\mathrm{Tr}\,[i\gamma_5\gamma_\mu\varrho(-p)]}{\mathrm{Tr}\,[\varrho(-p)]} = -4D_\mu. \tag{126}$$

Analogous to (109) we obtain

$$C_{\mu\nu} = \frac{1}{8im}(p_\mu \xi_\nu - \xi_\mu p_\nu). \tag{127}$$

Ultimately from (123)–(127) we find

$$\varrho(-p) = -\Lambda(-p)\frac{1+i\gamma_5\hat{\xi}}{2}. \tag{128}$$

For unpolarized antiparticles

$$\varrho_0(-p) = \tfrac{1}{2}\sum_s u^s(-p)\,\bar{u}^s(-p) = -\tfrac{1}{2}\Lambda(-p). \tag{129}$$

In conclusion we shall obtain the expression for the density matrix of the neutrino. Denote the spinors which describe the neutrino with definite helicity by

$$\left.\begin{aligned} u^r_+(\boldsymbol{p}) &= u^r(p), \\ u^r_-(-\boldsymbol{p}) &= u^r(-p). \end{aligned}\right\} \tag{130}$$

We normalize the spinors $u^r(p)$ and $u^r(-p)$ by the conditions (for $m = 0$, $\bar{u}(p)\,u(p) = 0$)

$$\left.\begin{aligned} (u^{r'}(p))^+\,u^r(p) &= \delta_{rr'}, \\ (u^{r'}(-p))^+\,u^r(-p) &= \delta_{r'r}. \end{aligned}\right\} \tag{131}$$

Note that for such a normalization the neutrino field operator is

$$\psi_\nu(x) = \frac{1}{(2\pi)^{3/2}}\int u^r(p)\,c_r(p)\,e^{ipx}\,d\boldsymbol{p} + \frac{1}{(2\pi)^{3/2}}\int u^r(-p)\,d_r^+(p)\,e^{-ipx}\,d\boldsymbol{p}. \tag{132}$$

In accordance with the present V–A theory of weak interactions the spinors describing the neutrino enter the matrix elements of the process in the form $(1+\gamma_5)\,u^r(p)$ (absorption of the neutrino) or $((1+\gamma_5)\,u^r(p))^+$ (emission of the neutrino). Since $\gamma_5 u^{(1)}(p) = -u^{(1)}(p)$ [see (33)] this means that the neutrino can be found only in the state with helicity of -1. Similarly it is easy to see that the helicity of the antineutrino is $+1$.

The density matrix of the neutrino is given consequently by the expression

$$\varrho_{\sigma\sigma'}(p) = u_\sigma^{(-1)}(p)\,\bar{u}_{\sigma'}^{(-1)}(p). \tag{133}$$

Let us expand this matrix into a complete set. We obtain

$$\varrho(p) = a + b_\mu\gamma_\mu + c_{\mu\nu}\sigma_{\mu\nu} + id_\mu\gamma_5\gamma_\mu + e\gamma_5. \tag{134}$$

The coefficient a is equal to

$$a = \tfrac{1}{4}\,\mathrm{Tr}\,\varrho = \tfrac{1}{4}\,\bar{u}^{(-1)}(p)\,u^{(-1)}(p) = 0. \tag{135}$$

For b_μ we find

$$b_\mu = \tfrac{1}{4}\,\mathrm{Tr}\,\varrho(p)\,\gamma_\mu = \tfrac{1}{4}\,\bar{u}^{(-1)}(p)\,\gamma_\mu u^{(-1)}(p). \tag{136}$$

This matrix element is not difficult to calculate. The spinors $u(p)$ and $u^+(p)$ satisfy the equations

$$\boldsymbol{\alpha p}\,u(p) = p_0 u(p), \quad u^+(p)\,\boldsymbol{\alpha p} = u^+(p)\,p_0. \tag{137}$$

Multiplying the first equation on the left by $u^{+}(p)\alpha_i$ and the second on the right by $\alpha_i u(p)$ and adding the resulting relationships, we find

$$u^{+}(p)\,u(p)\,p_i = ip_0\bar{u}(p)\,\gamma_i u(p). \tag{138}$$

From this we obtain

$$i\bar{u}(p)\,\gamma_\mu u(p) = \frac{p_\mu}{p_0}\,u^{+}(p)\,u(p). \tag{139}$$

In the normalization we have chosen [see (131)] the coefficient b_μ therefore is

$$b_\mu = \frac{1}{4ip_0}\,p_\mu. \tag{140}$$

Furthermore, for d_μ we obtain

$$d_\mu = \tfrac{1}{4}\,\mathrm{Tr}\,[\varrho(p)\,i\gamma_5\gamma_\mu] = \tfrac{1}{4}\bar{u}^{(-1)}(p)\,i\gamma_5\gamma_\mu u^{(-1)}(p). \tag{141}$$

Considering that $\gamma_5 u^{(-1)}(p) = u^{(-1)}(p)$ we find with the help of (139)

$$d_\mu = -\frac{1}{4p_0}\,p_\mu. \tag{142}$$

It is easy to see that the coefficients $c_{\mu\nu}$ and e are zero. Actually,

$$\begin{aligned} e &= \tfrac{1}{4}\,\mathrm{Tr}\,(\varrho\gamma_5) = \tfrac{1}{4}\bar{u}^{(-1)}(p)\,\gamma_5 u^{(-1)}(p) = \tfrac{1}{4}\bar{u}^{(-1)}(p)\,u^{(-1)}(p) = 0,\\ c_{\mu\nu} &= \tfrac{1}{8}\,\mathrm{Tr}\,(\varrho\sigma_{\mu\nu}) = \tfrac{1}{8}\bar{u}^{(-1)}(p)\,\sigma_{\mu\nu}u^{(-1)}(p)\\ &= -\tfrac{1}{8}\bar{u}^{(-1)}(p)\,\gamma_5\sigma_{\mu\nu}\gamma_5 u^{(-1)}(p) = -\tfrac{1}{8}\bar{u}^{(-1)}(p)\,\sigma_{\mu\nu}u^{(-1)}(p) = 0. \end{aligned} \tag{143}$$

We ultimately obtain for the neutrino density matrix

$$\varrho(p) = \tfrac{1}{2}(1+\gamma_5)\frac{\hat{p}}{2ip_0}\,. \tag{144}$$

The density matrix of the antineutrino is defined by

$$\varrho_{\sigma\sigma'}(-p) = u_\sigma^{(1)}(-p)\,\bar{u}_{\sigma'}^{(1)}(-p), \tag{145}$$

where

$$\Sigma(-\boldsymbol{n})\,u^{(1)}(-p) = u^{(1)}(-p). \tag{146}$$

With the help of (29) it is easy to show that

$$\gamma_5 u^{(1)}(-p) = u^{(1)}(-p). \tag{147}$$

Using this relationship we find from (145)

$$\varrho(-p) = \tfrac{1}{2}(1+\gamma_5)\frac{\hat{p}}{2ip_0}\,. \tag{148}$$

INDEX

OTHER TITLES IN THE SERIES IN NATURAL PHILOSOPHY

Vol. 47. NIETO—The Titius–Bode Law of Planetary Distances: Its History and Theory
Vol. 48. WAGNER—Introduction to the Theory of Magnetism
Vol. 49. IRVINE—Nuclear Structure Theory
Vol. 50. STROHMEIER—Variable Stars
Vol. 51. BATTEN—Binary and Multiple Systems of Stars
Vol. 52. ROUSSEAU and MATHIEU—Problems in Optics
Vol. 53. BOWLER—Nuclear Physics
Vol. 54. POMRANING—The Equations of Radiation Hydrodynamics
Vol. 55. BELINFANTE—A Survey of Hidden-variables Theories
Vol. 56. SCHEIBE—The Logical Analysis of Quantum Mechanics
Vol. 57. ROBINSON—Macroscopic Electromagnetism
Vol. 58. GOMBÁS and KISDI—Wave Mechanics and Its Applications
Vol. 59. KAPLAN and TSYTOVICH—Plasma Astrophysics
Vol. 60. KOVÁCS and ZSOLDOS—Dislocations and Plastic Deformation
Vol. 61. AUVRAY and FOURRIER—Problems in Electronics
Vol. 62. MATHIEU—Optics
Vol. 63. ATWATER—Introduction to General Relativity
Vol. 64. MÜLLER—Quantum Mechanics: A Physical World Picture